VOID

Library of
Davidson College

THE HEAT EQUATION

Pure and Applied Mathematics

A Series of Monographs and Textbooks

Editors **Samuel Eilenberg and Hyman Bass**

Columbia University, New York

RECENT TITLES

XIA DAO-XING. Measure and Integration Theory of Infinite-Dimensional Spaces: Abstract Harmonic Analysis

RONALD G. DOUGLAS. Banach Algebra Techniques in Operator Theory

WILLARD MILLER, JR. Symmetry Groups and Their Applications

ARTHUR A. SAGLE AND RALPH E. WALDE. Introduction to Lie Groups and Lie Algebras

T. BENNY RUSHING. Topological Embeddings

JAMES W. VICK. Homology Theory: An Introduction to Algebraic Topology

E. R. KOLCHIN. Differential Algebra and Algebraic Groups

GERALD J. JANUSZ. Algebraic Number Fields

A. S. B. HOLLAND. Introduction to the Theory of Entire Functions

WAYNE ROBERTS AND DALE VARBERG. Convex Functions

A. M. OSTROWSKI. Solution of Equations in Euclidean and Banach Spaces, Third Edition of Solution of Equations and Systems of Equations

H. M. EDWARDS. Riemann's Zeta Function

SAMUEL EILENBERG. Automata, Languages, and Machines: Volume A. *In preparation:* Volume B

MORRIS HIRSCH AND STEPHEN SMALE. Differential Equations, Dynamical Systems, and Linear Algebra

WILHELM MAGNUS. Noneuclidean Tesselations and Their Groups

J. DIEUDONNÉ. Treatise on Analysis, Volume IV

FRANÇOIS TREVES. Basic Linear Partial Differential Equations

WILLIAM M. BOOTHBY. An Introduction to Differentiable Manifolds and Riemannian Geometry

BRAYTON GRAY. Homotopy Theory: An Introduction to Algebraic Topology

ROBERT A. ADAMS. Sobolev Spaces

JOHN J. BENEDETTO. Spectral Synthesis

D. V. WIDDER. The Heat Equation

In preparation

IRVING E. SEGAL. Mathematical Cosmology and Extragalactic Astronomy

WERNER GREUB, STEPHEN HALPERIN, AND RAY VANSTONE. Connections, Curvature, and Cohomology: Volume III, Cohomology of Principal Bundles and Homogeneous Spaces

J. DIEUDONNÉ. Treatise on Analysis, Volume II, enlarged and corrected printing

I. MARTIN ISAACS. Character Theory of Finite Groups

THE HEAT EQUATION

D. V. WIDDER

Department of Mathematics
Harvard University
Cambridge, Massachusetts

ACADEMIC PRESS   New York San Francisco London   1975

A Subsidiary of Harcourt Brace Jovanovich, Publishers

COPYRIGHT © 1975, BY ACADEMIC PRESS, INC.
ALL RIGHTS RESERVED.
NO PART OF THIS PUBLICATION MAY BE REPRODUCED OR
TRANSMITTED IN ANY FORM OR BY ANY MEANS, ELECTRONIC
OR MECHANICAL, INCLUDING PHOTOCOPY, RECORDING, OR ANY
INFORMATION STORAGE AND RETRIEVAL SYSTEM, WITHOUT
PERMISSION IN WRITING FROM THE PUBLISHER.

ACADEMIC PRESS, INC.
111 Fifth Avenue, New York, New York 10003

United Kingdom Edition published by
ACADEMIC PRESS, INC. (LONDON) LTD.
24/28 Oval Road, London NW1

Library of Congress Cataloging in Publication Data

Widder, David Vernon, (date)
    The heat equation.

    (Pure and applied mathematics; a series of monographs
and textbooks)
    Bibliography: p.
    Includes index.
    1. Heat equation.   I. Title.   II. Series.
QA3.P8 [QA377]   510'.8s [515'.353]   74-30816
ISBN 0–12–748540–6

AMS (MOS) 1970 Subject Classifications: 35K05, 44A15, 80A20

PRINTED IN THE UNITED STATES OF AMERICA

To my son David

CONTENTS

Preface   xi
Symbols and Notation   xiii

Chapter I Introduction   1

    1 Introduction   1
    2 The Physical Model   1
    3 The Heat Equation   3
    4 Generalities   5
    5 Basic Solutions of the Heat Equation   8
    6 Methods of Generating Solutions   10
    7 Definitions and Notations   14

Chapter II Boundary-Value Problems   17

    1 Introduction   17
    2 Uniqueness   18
    3 The Maximum Principle   20
    4 A Criterion for Temperature Functions   23
    5 Solution of Problem I in a Special Case   24
    6 Uniqueness for the Infinite Rod   26

Chapter III Further Developments — 30

1. Introduction 30
2. The Source Solution 30
3. The Addition Formula for $k(x, t)$ 32
4. The Homogeneity of $k(x, t)$ 34
5. An Integral Representation of $k(x, t)$ 35
6. A Further Addition Formula for $k(x, t)$ 37
7. Laplace Transform of $k(x, t)$ 38
8. Laplace Transform of $h(x, t)$ 39
9. Operational Calculus 41
10. Three Classes of Functions 44
11. Examples of Class II 46
12. Relation among the Classes 49
13. Series Expansions of Functions in Class I 50
14. Series Expansions of Functions in Class II 52
15. Series Expansions of Functions in Class III 53
16. A Temperature Function Which Is Not Entire in the Space Variable 58

Chapter IV Integral Transforms — 60

1. Poisson Transforms 60
2. Convergence 62
3. Poisson Transform in H 64
4. Analyticity 64
5. Inversion of the Poisson–Lebesgue Transform 65
6. Inversion of the Poisson–Stieltjes Transform 68
7. The h-Transform 70
8. h-Transform in H 74
9. Analyticity 75
10. Inversion of the h-Lebesgue Transform 78
11. The k-Transform 80
12. A Basic Integral Representation 82
13. Analytic Character of Every Temperature Function 84

Chapter V Theta-Functions — 86

1. Introduction 86
2. Analyticity 88
3. θ-Functions in H 89
4. Alternate Expansions 90
5. Two Positive Kernels 92
6. A θ-Transform 94
7. A φ-Transform 97
8. Fourier's Ring 100
9. A Solution of the First Boundary-Value Problem 101
10. Uniqueness 102

Chapter VI Green's Function — 107

1. Green's Function for a Rectangle 107
2. An Integral Representation 109

    3 Problem I Again   110
    4 A Property of $G(x, t; \xi, n)$ 112
    5 Green's Function for an Arbitrary Rectangle   113
    6 Series of Temperature Functions   114
    7 The Reflection Principle   115
    8 Isolated Singularities   116

Chapter VII Bounded Temperature Functions — 122

    1 The Infinite Rod   122
    2 The Semi-Infinite Rod   124
    3 Semi-Infinite Rod, Continued   126
    4 Semi-Infinite Rod, General Case   127
    5 The Finite Rod   130

Chapter VIII Positive Temperature Functions — 132

    1 The Infinite Rod   132
    2 Uniqueness, Positive Temperatures on an Infinite Rod   133
    3 Stieltjes Integral Representation, Infinite Rod   136
    4 Uniqueness, Semi-Infinite Rod   137
    5 Representation, Semi-Infinite Rod   140
    6 The Finite Rod   147
    7 Examples   151
    8 Further Classes of Temperature Functions   153

Chapter IX The Huygens Property — 155

    1 Introduction   155
    2 Blackman's Example   159
    3 Conditionally Convergent Poisson Integrals   161
    5 Heat Polynomials and Associated Functions   165

Chapter X Series Expansions of Temperature Functions — 169

    1 Introduction   169
    2 Asymptotic Estimates   171
    3 A Generating Function   175
    4 Region of Convergence   177
    5 Strip of Convergence   180
    6 Representation by Series of Heat Polynomials   183
    7 The Growth of an Entire Function   185
    8 Expansions in Series of Associated Functions   186
    9 A Further Criterion   188
    10 Examples   191

Chapter XI Analogies — 195

    1 Introduction   195
    2 The Appell Transformation   197
    3 Heat Polynomials   197
    4 Associated Functions   198

5 The Huygens Property   198
   6 The Operators e^{cD} and e^{cD^2} 199
   7 Biorthogonality   199
   8 Generating Functions   200
   9 Polynomial Expansions   200
  10 Associated Function Expansions   200
  11 Criteria for Polynomial Expansions   201
  12 Criteria for Expansions in Series of Associated Functions   201

Chapter XII Higher Dimensions 204

   1 Introduction   204
   2 The Heat Equation for Solids   205
   3 Notations and Definitions   207
   4 Generating Functions   209
   5 Expansions in Series of Polynomials   210
   6 An Example   214

Chapter XIII Homogeneous Temperature Functions 216

   1 Introduction   216
   2 The Totality of Homogeneous Temperature Functions   218
   3 Recurrence Relations   222
   4 Continued Fraction Developments   223
   5 Decomposition of the Basic Functions   226
   6 Summary   227
   7 Series of Polynomials   227
   8 First Kind, Negative Degree   230
   9 Second Kind, Positive Degree   231
  10 Second Kind, Negative Degree   232
  11 Examples   232

Chapter XIV Miscellaneous Topics 235

   1 Positive Temperature Functions   235
   2 Positive Definite Functions   237
   3 Positive Temperature Functions, Concluded   239
   4 A Statistical Problem   241
   5 Examples   244
   6 Statistical Problem Concluded   246
   7 Alternate Inversion of the h-Transform   249
   8 Time-Variable Singularities   253

Bibliography   259
Index   263

PREFACE

This book is designed for students who have had no previous knowledge of the theory of heat conduction nor indeed of the general theory of partial differential equations. On the other hand, a degree of mathematical sophistication is assumed, in that the reader is expected to be familiar with the basic results of the theory of functions of a complex variable, Laplace transform theory, and the standard working tools involving Lebesgue integration. It should be understandable to beginning graduate students or to advanced undergraduates.

The heat equation is derived in Chapters I and XII as a consequence of two basic postulates, easily accepted from physical experience. From this point on, the theorems and results are logical consequences of the heat equation. If the conclusions are at variance with physical facts, and they are slightly so, the fault must be traced to the postulates. For example, the equation forces the conclusion that "action at a distance" is possible. That is, heat introduced at any point of a linear bar raises temperature *instantaneously* at remote portions of the bar. This scandalizes reason and contradicts experiment, so that we must conclude that the postulates are only approximations to the physical situation. But it has also been evident since Fourier's time that they are *good* approximations.

The early chapters develop a theory of the integral transforms that are needed for the integral representations of solutions of the heat equation.

Results that are needed here about the theta-functions of Jacobi are proved in Chapter V. Transforms for which theta-functions are kernels are used for solving boundary-value problems for the finite bar. No previous knowledge of theta-functions is assumed. At only one point is an unproved formula about them employed, and even here, a second approach to the desired result avoids use of that one formula.

Much of the material in Chapters VIII–XIV is based on the author's own research, but it is presented in simplified form. The emphasis is on the expansion of solutions of the heat equation into infinite series. Here the analogies from complex analysis of series developments of analytic functions are very revealing. These are pointed out in detail in Chapter XI. In the final chapter the essential results from four research papers are given simplified proof.

All the material could probably be presented in a half course. More realistically, Chapters V, VI, XII, XIII, and those parts of Chapters VII and VIII dealing with the finite rod, could be omitted. These could be replaced by classic boundary-value problems.

Theorems are generally stated in the same systematic and compact style used by the author in "Advanced Calculus" and in "An Introduction to Transform Theory." The few logical symbols needed to accomplish this are for the most part self-explanatory, but a few are explained parenthetically when introduced.

SYMBOLS AND NOTATION

Page	Symbol	Meaning
7	$\Rightarrow$	implies
7	$\in$	is a member of
7	$(X\|P)$	elements X having property P
7	C^n	continuous with derivatives of order $\leq n$
8	C	continuous
8	$v_n(x, t)$	heat polynomial
9	[ ]	largest integer $\leq$
10	$\rightarrow$	is transformed into
10	$k(x, t)$	source solution
10	$h(x, t)$	derived source solution
10	$l(x, t)$	complementary error function
10	$\theta(x, t)$	theta-function
10	$\varphi(x, t)$	derived theta-function
13, 32	$*$	convolution
14	H	satisfies the heat equation is a temperature function
14	H^*	satisfies the adjoint heat equation
14	$\Leftrightarrow$	implies and is implied by if and only if

SYMBOLS AND NOTATION

Page	Symbol	Meaning
16	′ (prime)	derivative with respect to *first* variable
19	→	approaches
25	$O(\ )$	is of the order of
25, 36	$\ll$	is dominated by
26	A	analytic
41	e^{aD}	translation operator
42	e^{aD^2}	Poisson transform operator
43	$\cosh a\sqrt{\mathcal{D}}$	operator on time variable
43, 250	$e^{a\sqrt{\mathcal{D}}}$	operator on time variable
44	I	class of functions (real variable)
44	II	class of functions (real variable)
45	III	class of functions (real variable)
60	L	Lebesgue integrable
60	V	bounded variation
68	$o(\ )$	is less than the order of
71	↑	nondecreasing
72	↓	nonincreasing
107	R	a rectangle
107	Γ	boundary of R
107	γ	sides and base of R
108	γ^*	sides and top of R
122	B	bounded
156	H^Δ	difference of positive functions of H
156	H^o	Huygens property
170	H_n	Hermite polynomial
170	$\delta_{m,n}$	Kronecker delta
185	$\{m, n\}$	growth of an entire function m, the maximum order; n, the maximum type
198	A^o	analytic (restricted)
203	Ap[ ]	Appell transformation
216	$V_n(x, t)$	homogeneous of first kind
217	$h_n(x, t)$	homogeneous of second kind
217	$H_n(x, t)$	homogeneous of second kind
233	$\binom{m}{n}$	binomial coefficient
238	PD	positive definite
247	$\hat{f}$	Fourier transform of f
247	P_a	subclass of frequency functions
250	$f^{(-1/2)}$	Riemann–Liouville fractional integral
251	$\sqrt{\mathcal{D}} = \mathcal{D}^{-1/2}$	Riemann–Liouville fractional integral

THE HEAT EQUATION

Chapter I

INTRODUCTION

1 INTRODUCTION

In this chapter we shall set down our basic assumptions behind the theory of heat conduction and then derive the heat equation in two dimensions therefrom. We point out its relation to the general theory of partial differential equations and establish a basic Green's formula. We give many examples of solutions of the equation and discuss methods of obtaining new ones. Finally, we set the stage for deeper study by introducing formal definitions and notations.

2 THE PHYSICAL MODEL

In the study of heat, a physical model may be imagined in which heat is considered to be a fluid inside matter, free to flow from one position to another. The amount of fluid present is measured in some unit such as the calorie (cal) or BTU (British Thermal Unit). Evidence of its presence in matter is the temperature thereof, it being assumed that the more heat present the higher the temperature, and that it flows from places of high temperature to places of low temperature. Temperature can be measured directly by a thermometer; the quantity of heat present is inferred indirectly, as we see by the following definition.

Definition 2 A unit of heat is the amount needed to raise the temperature of one unit of water by one unit of temperature.

For example, in cgs units the unit is called the *calorie* and is the amount of heat necessary to raise one gram of water one degree centigrade. A BTU is the amount to raise one pound one degree Fahrenheit. This is a much larger unit since 1 BTU = 252 cal, approximately.

This definition already makes an assumption about the behavior of heat, for it implies that temperature rise caused by a given quantity of heat is independent of the starting temperature. By experiment this is found to be only approximately true. But in setting up our model we take cognizance of the fact that we are making *approximations* to actual physical conditions.

We now make two postulates about the nature of heat, both of which can be roughly verified by experiment.

Postulate A (Absorption) The amount of increase in the quantity of heat ΔQ in a material is directly proportional to the mass m of the material and to the increase in temperature Δu:

$$\Delta Q = cm \, \Delta u.$$

Here c is the constant of proportionality and is assumed to vary only with the material. It is called the *specific heat* of the material. From Definition 2, $c = 1$ for water. For lead $c = 0.03$ and for silver $c = 0.06$, approximately.

Postulate B (Conduction) Consider a straight bar of homogeneous material, sides insulated, of length Δx and of constant cross section A. If the two ends are held constantly at two different temperatures differing by Δu, the temperature along the bar will vary linearly, and the amount of flow will depend on the abruptness in the change of temperature $\Delta u/\Delta x$. Further, this quantity ΔQ is directly proportional to A, to $\Delta u/\Delta x$, and to Δt, the amount of time elapsed:

$$\Delta Q = -lA \frac{\Delta u}{\Delta x} \Delta t.$$

Here l is the constant of proportionality, called the *thermal conductivity* of the material, and the flow is in the direction of increasing x when $\Delta u/\Delta x < 0$. To make the postulate graphic, we may think of the bar as extended along the x-axis of an x,u-plane as in Figure 1.

3. THE HEAT EQUATION

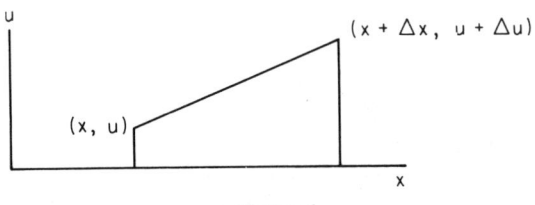

Figure 1

If Δx and Δu are positive, as in the figure, then the flow is opposite to the direction of increasing x and its rate of flow $\Delta Q/\Delta t$ is taken to be negative, $-lA\,\Delta u/\Delta x$. In particular, if the temperature gradient $\Delta u/\Delta x$ is 1° centigrade per centimeter, then l is the number of calories of heat flowing across 1 sq cm of cross section in 1 sec. For water $l = 0.0014$, for lead $l = 0.083$, and for silver $l = 1.0006$, approximately.

If the increase in temperature in a given segment of matter of mass m is a function of time, Postulate A shows that

$$\left.\frac{\partial Q}{\partial t}\right|_{t_0} = cm \left.\frac{\partial u}{\partial t}\right|_{t_0}. \tag{1}$$

That is, the instantaneous rate of increase, at time t_0, of the quantity of heat in the segment is proportional to the instantaneous rate of rise in temperature there.

If the temperature in the above bar is a function of x on the given interval, Postulate B shows that

$$\left.\frac{\partial Q}{\partial t}\right|_{x_0} = -lA \left.\frac{\partial u}{\partial x}\right|_{x_0} \tag{2}$$

at a given instant t_0. That is, the rate of flow across the surface $x = x_0$ is proportional to the temperature gradient there. As pointed out above, the sign of $\partial u/\partial x$ determines the direction of flow (to the left in Figure 1 if $\partial u/\partial x > 0$).

3 THE HEAT EQUATION

The heat equation in two dimensions is the partial differential equation

$$\frac{\partial^2 u}{\partial x^2} = \frac{\partial u}{\partial t}. \tag{1}$$

Its generalization to higher dimensions is

$$\frac{\partial^2 u}{\partial x_1^2} + \frac{\partial^2 u}{\partial x_2^2} + \cdots + \frac{\partial^2 u}{\partial x_n^2} = \frac{\partial u}{\partial t}.$$

It has applications in various branches of science, one of the most important of which is in the theory of heat conduction, as the name implies. Our chief interest will be in the analytic consequences of the equation, but a physical interpretation of some of our results will be a useful guide to the analysis. Accordingly, we shall begin with a derivation of equation (1) as a consequence of Postulates A and B.

We consider the homogeneous bar of the previous section placed on an x-axis as there described. Denote its temperature at a point x of the bar and at time t by $u(x, t)$. We show that $u(x, t)$ must satisfy (1) if Postulates A and B hold. Assume uniform density ρ.

Isolate a segment of the bar between points x_0 and $x_0 + \Delta x$. From 2(1) we see that the rate of absorption $(\partial Q/\partial t)(x, t_0)$ at a given point x and at time t_0 is proportional to $(\partial u/\partial t)(x, t_0)$. The average value of the latter function over the segment will occur at some interior point $x_0 + \theta \Delta x$, so that the rate of absorption of heat in the whole segment will be

$$\left.\frac{\partial Q}{\partial t}\right|_{t=t_0} = c\rho \, \Delta x \, A \, \frac{\partial u}{\partial t}(x_0 + \theta \Delta x, t_0), \qquad 0 < \theta < 1. \qquad (2)$$

By the conduction postulate we may now compute this rate in another way, assuming that what is absorbed must be flowing in. From 2(2), the rate at which heat is entering the segment at time t_0, through the surface $x = x_0$, is

$$- lA \frac{\partial u}{\partial x}(x_0, t_0),$$

while the rate at which it is leaving through $x = x_0 + \Delta x$ is

$$- lA \frac{\partial u}{\partial x}(x_0 + \Delta x, t_0),$$

so that the rate of accumulation (and hence absorption) is

$$lA\left[\frac{\partial u}{\partial x}(x_0 + \Delta x, t_0) - \frac{\partial u}{\partial x}(x_0, t_0)\right] = lA \frac{\partial^2 u}{\partial x^2}(x_0 + \theta' \Delta x, t_0) \, \Delta x. \qquad (3)$$

Here we have used the law of the mean, θ' being a suitable number between 0 and 1. Equating (2) and (3) and allowing Δx to approach zero, we obtain

$$l \frac{\partial^2 u}{\partial x^2}(x_0, t_0) = c\rho \frac{\partial u}{\partial t}(x_0, t_0). \tag{4}$$

The constant $a = l/(c\rho)$ is called the *thermometric conductivity* (Maxwell) or the *diffusivity* (Lord Kelvin). By its use, equation (4) becomes

$$a \frac{\partial^2 u}{\partial x^2} = \frac{\partial u}{\partial t}. \tag{5}$$

The introduction of a new time variable $t' = at$ reduces equation (5) to (1), and we shall assume henceforth that this is done. But it must be remembered that if practical application of a solution $u(x, t)$ of (1) is to be made, it is $u(x, at)$ that is a solution of (5). We summarize our result in theorem form.

Theorem 3 The temperature $u(x, t)$ at a point of coordinate x and at time t of a homogeneous insulated bar (as described in §2) satisfies the equation

$$\frac{\partial u}{\partial t} = \frac{l}{c\rho} \frac{\partial^2 u}{\partial x^2} = a \frac{\partial^2 u}{\partial x^2}.$$

Here l is the conductivity, c the specific heat, ρ the density, and a the diffusivity of the bar.

The absorption postulate involves not the absolute quantity of heat in a body but rather a change in the quantity. Usually we consider the amount present as the increase in the quantity from $0°K$. For a given temperature distribution $u(x, t)$ in a bar, we can compute by integration the total amount of heat in a segment (a, b) thereof. At an instant t_0 the increase in temperature from zero at a point x_0 is $u(x_0, t_0)$. Then by Postulate A the total quantity of heat in the segment (a, b) at time t_0 is

$$Q(t_0) = c\rho \int_a^b u(x, t_0)\, dx. \tag{6}$$

4 GENERALITIES

The heat equation is a very special case of the general linear second-order partial differential equation:

$$L[u] = A \frac{\partial^2 u}{\partial x^2} + B \frac{\partial^2 u}{\partial x\, \partial t} + C \frac{\partial^2 u}{\partial t^2} + D \frac{\partial u}{\partial x} + E \frac{\partial u}{\partial t} + Fu = 0. \tag{1}$$

Here the coefficients may be functions of x and t but not of u. The equation is classified into types as follows:

$B^2 - 4AC < 0$ (elliptic); $\quad \dfrac{\partial^2 u}{\partial x^2} + \dfrac{\partial^2 u}{\partial t^2} = 0$ (Laplace);

$B^2 - 4AC > 0$ (hyperbolic); $\quad \dfrac{\partial^2 u}{\partial x^2} - \dfrac{\partial^2 u}{\partial t^2} = 0$ (wave);

$B^2 - 4AC = 0$ (parabolic); $\quad \dfrac{\partial^2 u}{\partial x^2} - \dfrac{\partial u}{\partial t} = 0$ (heat).

On the right we have given the simplest and most familiar illustrations of the corresponding types. The names adopted come of course from analytic geometry, the analogy of equation (1) with that for the general conic being apparent.

Since the heat equation is such a very special case of (1) we make no use of the general theory here. We seek only to show the position of the heat equation in its setting as a particular case of a larger theory. With this in view we point out that the *characteristic curves* or *characteristics* associated with equation (1) are defined by the differential equation

$$A\, dt^2 - B\, dx\, dt + C\, dx^2 = 0. \tag{2}$$

These curves are important in view of their role in the solution of boundary value problems for (1). The so-called *Cauchy problem* is the determination of a solution $u(x, t)$ of (1) such that

$$u(x, t) = \varphi(x, t), \quad u_x(x, t) = \psi(x, t)$$

for prescribed functions φ and ψ and along some prescribed curve $\omega(x, t) = 0$. A general existence theorem states that this is possible if the latter curve is *not* a characteristic. For the heat equation $A = 1$, $B = C = 0$, and equation (2) reduces to $dt^2 = 0$. That is, any line parallel to the x-axis is characteristic. In light of the existence theorem we should not expect to find a solution $u(x, t)$ of the heat equation satisfying

$$u(x, 0) = \varphi(x), \quad u_x(x, 0) = \psi(x) \tag{3}$$

for arbitrary functions $\varphi(x)$ and $\psi(x)$, since the line $t = 0$ is a characteristic. In fact we shall show later that the first of equations (3) above is generally sufficient to determine a solution of the heat equation. In this case of a boundary-value problem involving a single prescribed function

4. GENERALITIES

on a characteristic, the term Cauchy problem is still customarily used. Accordingly, we refer to either of the following as

Cauchy's Problem Find $u(x, t)$ satisfying $u_{xx} = u_t$ such that

$$\text{A.} \quad u(x, 0) = \varphi(x)$$

or

$$\text{B.} \quad u(0, t) = \varphi(t), \quad u_x(0, t) = \psi(t).$$

In case A the prescribed boundary curve is the characteristic $t = 0$; in case B we have chosen it as the noncharacteristic $x = 0$. Of course, other boundary curves may be needed for particular problems, but those chosen are the ones most frequently met in practice.

In the general theory, it is useful to define an equation *adjoint* to (1):

$$L^*[u] = \frac{\partial^2}{\partial x^2}(Au) + \frac{\partial^2(Bu)}{\partial x\, \partial t} + \frac{\partial^2(Cu)}{\partial t^2}$$

$$- \frac{\partial}{\partial x}(Du) - \frac{\partial}{\partial t}(Eu) + Fu = 0.$$

It is especially useful in the classical "Green's theorem," which we need not state here in full generality since we shall state and prove it independently as it applies to the heat equation. For that equation

$$L[u] = \frac{\partial^2 u}{\partial x^2} - \frac{\partial u}{\partial t}, \quad L^*[u] = \frac{\partial^2 u}{\partial x^2} + \frac{\partial u}{\partial t}.$$

Let R be the following rectangle with sides parallel to the axes

$$R = \{(x, t) \mid a \leq x \leq b, \ c \leq t \leq d\},$$

and denote its boundary by Γ. [We use here a familiar notation $\{X \mid P\}$ for the set of all elements X having property P.] With these notations we now state Green's theorem.

Theorem 4

1. $u(x, t), v(x, t) \in C^2$ in R

$$\Rightarrow \iint_R (vL[u] - uL^*[v])\, dx\, dt = \int_\Gamma uv\, dx + (vu_x - uv_x)\, dt, \quad (4)$$

where integration of the line integral is counterclockwise over the boundary Γ of the rectangle R.

The notation of the hypothesis means that u and v are continuous with their derivatives of orders 1 and 2 in and on the boundary of R. To prove the theorem we appeal to the familiar relation

$$\iint_R (Q_x - P_t)\, dx\, dt = \int_\Gamma P\, dx + Q\, dt, \tag{5}$$

where

$$P = uv, \qquad Q = vu_x - uv_x.$$

Then

$$Q_x - P_t = v(u_{xx} - u_t) - u(v_{xx} + v_t) = vL[u] - uL^*[v],$$

and equation (5) reduces to (4).

It is clear that the theorem could be extended to any region R bounded by Γ for which (5) is valid. But the rectangle is the useful region for our purposes. We shall refer to Theorem 4 as Green's theorem. Since (5) requires only that $P, Q \in C^1$, hypothesis 1 could be relaxed to read $u, v \in C^1$ and $u_{xx}, v_{xx} \in C$. Thus Green's theorem is applicable to solutions of the heat equation or its adjoint.

5 BASIC SOLUTIONS OF THE HEAT EQUATION

We list here a few of the most important solutions of the heat equation.

A Exponential solutions If constants α and β are chosen so as to make $\exp(\alpha x + \beta t)$ a solution of the heat equation, $\beta = \alpha^2$. We thus obtain a one parameter family of solutions

$$u(x, t) = e^{\alpha x + \alpha^2 t}. \tag{1}$$

Since α may be real or complex, the functions

$$e^{-\alpha^2 t} \cos \alpha x, \qquad e^{-\alpha^2 t} \sin \alpha x, \qquad e^{\alpha^2 t} \cosh \alpha x, \qquad e^{\alpha^2 t} \sinh \alpha x \tag{2}$$

are immediately seen to be solutions also.

B Heat polynomials We define the *heat polynomial* of degree n as the coefficient of $z^n/n!$ in the expansion

$$e^{xz + tz^2} = \sum_{n=0}^{\infty} v_n(x, t) \frac{z^n}{n!}. \tag{3}$$

Since the exponential on the left is in the family (1) we would expect the

5. BASIC SOLUTIONS OF THE HEAT EQUATION

polynomials $v_n(x, t)$ to be solutions also. We show this directly by differentiating equation (3) with respect to x and t.

$$e^{xz+tz^2}z = \sum_{n=0}^{\infty} \frac{\partial}{\partial x} v_n(x, t) \frac{z^n}{n!} = \sum_{n=0}^{\infty} v_n(x, t) \frac{z^{n+1}}{n!} ;$$

$$e^{xz+tz^2}z^2 = \sum_{n=0}^{\infty} \frac{\partial}{\partial t} v_n(x, t) \frac{z^n}{n!} = \sum_{n=0}^{\infty} v_n(x, t) \frac{z^{n+2}}{n!} .$$

By comparing coefficients we obtain

$$\frac{\partial}{\partial x} v_n(x, t) = n v_{n-1}(x, t), \quad n = 1, 2, \ldots,$$

$$\frac{\partial}{\partial t} v_n(x, t) = n(n-1) v_{n-2}(x, t) = \frac{\partial^2}{\partial x^2} v_n(x, t), \quad n = 2, 3, \ldots.$$

It is clear by inspection that $v_0 = 1$, $v_1 = x$ also satisfy the heat equation, so that all the heat polynomials are solutions.

We can obtain an explicit expression for v_n by use of Cauchy's rule for multiplying power series:

$$e^{xz} = \sum_{n=0}^{\infty} a_n z^n, \quad a_n = \frac{x^n}{n!} ;$$

$$e^{tz^2} = \sum_{n=0}^{\infty} b_n z^n, \quad b_{2n} = \frac{t^n}{n!}, \quad b_{2n+1} = 0;$$

$$e^{xz+tz^2} = \sum_{n=0}^{\infty} c_n z^n, \quad c_n = \sum_{k=0}^{n} a_k b_{n-k}.$$

Thus

$$v_n(x, t) = n! \, c_n = n! \sum_{k=0}^{[n/2]} \frac{t^k}{k!} \frac{x^{n-2k}}{(n-2k)!} ,$$

where $[n/2]$ means the largest integer $\leq n/2$. For example,

$$v_2 = x^2 + 2t, \quad v_3 = x^3 + 6xt,$$

$$v_4 = x^4 + 12x^2 t + 12t^2, \quad v_5 = x^5 + 20x^3 t + 60xt^2.$$

In particular, we have for $n = 0, 1, 2, \ldots$

$$v_n(x, 0) = x^n, \quad v_{2n}(0, t) = \frac{(2n)! \, t^n}{n!}, \quad v_{2n+1}(0, t) = 0.$$

C The source solution A very important solution is

$$k(x, t) = \frac{e^{-x^2/4t}}{\sqrt{4\pi t}}, \qquad t > 0.$$

We shall discuss its properties later and justify its name. To show at once that it is a solution use logarithmic differentiation:

$$\frac{k_x}{k} = -\frac{x}{2t}, \qquad \frac{k_{xx}}{k} = -\frac{1}{2t} + \frac{x^2}{4t} = \frac{k_t}{k}. \tag{4}$$

Hence $k_{xx} = k_t$, as asserted.

D The derived source solution We shall have frequent use for the function

$$h(x, t) = \frac{xk(x, t)}{t} = -2k_x(x, t), \qquad t > 0,$$

$$h_{xx} - h_t = -2\frac{\partial}{\partial x}(k_{xx} - k_t) = 0.$$

E Complementary error function Another solution is

$$l(x, t) = \operatorname{erfc}\left(\frac{x}{\sqrt{4t}}\right) = \frac{2}{\sqrt{\pi}} \int_{x/\sqrt{4t}}^{\infty} e^{-y^2} dy.$$

Clearly

$$l_x = -2k, \qquad l_{xx} = l_t = h.$$

F Elliptic theta functions Basic in later study will be two other solutions defined as sums of source solutions and derived source solutions:

$$\theta(x, t) = \sum_{n=-\infty}^{\infty} k(x + 2n\pi, t), \qquad t > 0; \tag{5}$$

$$\varphi(x, t) = \sum_{n=-\infty}^{\infty} h(x + 2n\pi, t), \qquad t > 0. \tag{6}$$

We shall show later that these series converge for all x and that their sums satisfy the heat equation. It is clear at once that θ and φ are periodic in x, and therein lies part of their usefulness.

6 METHODS OF GENERATING SOLUTIONS

We list here a variety of transformations which carry solutions into solutions. In what follows the arrow will stand for the transformation.

A Multiplication by a constant

$$u(x, t) \to cu(x, t).$$

B Addition

$$u_1(x, t), u_2(x, t) \to u_1(x, t) + u_2(x, t).$$

These methods are of course consequences of the linearity and homogeneity of the heat equation. We used the method ad infinitum to define $\theta(x, t)$ and $\varphi(x, t)$ in §5. By adding solutions 5(2) we would expect

$$\frac{1}{2\pi} + \frac{1}{\pi} \sum_{n=1}^{\infty} e^{-tn^2} \cos nx, \qquad t > 0,$$

$$\frac{2}{\pi} \sum_{n=1}^{\infty} n e^{-tn^2} \sin nx, \qquad t > 0,$$

to be solutions, if the series converge. We show later that their sums are $\theta(x, t)$ and $\varphi(x, t)$, respectively.

C Integration with respect to a parameter

$$u(x, t, y) \to \int_a^b u(x, t, y)\, dy.$$

This is the continuous analogue of method B. Again questions of convergence may be involved if the integral is improper. An example, which we shall establish later, involves the source solution

$$k(x, t) = \frac{1}{\pi} \int_0^{\infty} e^{-ty^2} \cos xy\, dy, \qquad t > 0.$$

Here we have again used the family of solutions 5(2).

D Differentiation with respect to a parameter

$$u(x, t, y) \to \frac{\partial}{\partial y} u(x, t, y).$$

This again is an extension of methods A and B since

$$\frac{\partial u}{\partial y}(x, t, y) = \lim_{\Delta y \to 0} \frac{u(x, t, y + \Delta y) - u(x, t, y)}{\Delta y}.$$

For example,

$$\frac{\partial}{\partial y} e^{-y^2 t} \cos xy = e^{-y^2 t}(-2yt \cos xy - x \sin xy), \qquad t > 0,$$

E Translation

$$u(x, t) \to u(x - a, t - b).$$

This trivial transformation is none the less very useful. By combining it with methods A and C we would expect

$$\int_a^b k(x - y, t)g(y)\, dy, \qquad \int_a^b h(x, t - y)g(y)\, dy$$

to be solutions for arbitrary functions $g(y)$. In Chapter IV we shall discuss in detail integrals of this type, called convolutions.

F Affine transformation

$$u(x, t) \to u(ax, a^2 t).$$

It is easy to show that this is the most general affine transformation which carries one solution into another. For example, the solutions $e^t \cosh x$ and $e^{-t} \cos x$ are related by this method, with $a = i = \sqrt{-1}$.

G Differentiation with respect to x or t

$$u(x, t) \to u_x(x, t), \qquad u(x, t) \to u_t(x, t).$$

This is an obvious consequence of the fact that the heat equation is linear and homogeneous. We have already noted this result when observing that $h(x, t)$ and $\varphi(x, t)$ are solutions obtained as derivatives of $k(x, t)$ and $\theta(x, t)$, respectively.

H Integration with respect to x or t

$$u(x, t) \to v(x, t) = \int_a^x u(y, t)\, dy, \qquad \text{provided} \quad u_x(a, t) \equiv 0; \quad (1)$$

$$u(x, t) \to w(x, t) = \int_a^t u(x, y)\, dy, \qquad \text{provided} \quad u(x, a) \equiv 0. \quad (2)$$

This follows since

$$v_{xx} = u_x, \qquad v_t = \int_a^x u_t(y, t)\, dy = \int_a^x u_{xx}(y, t)\, dy = u_x(x, t) - u_x(a, t);$$

$$w_t = u, \qquad w_{xx} = \int_a^t u_{xx}(x, y)\, dy = \int_a^t u_y(x, y)\, dy = u(x, t) - u(x, a).$$

6. METHODS OF GENERATING SOLUTIONS

An example of (1) is

$$\int_0^x v_4(y, t)\, dy = \int_0^x (y^4 + 12y^2 t + 12t^2)\, dy = \frac{v_5(x, t)}{5}.$$

An example of (2) is

$$\int_0^t h(x, y)\, dy = l(x, t).$$

Here $l(x, t)$ is the function of Example E, §5. Note that $h(x, 0) \equiv 0$.

I Convolution

$$u(x, t) \to v(x, t) = \int_0^t u(x, t - y) g(y)\, dy, \qquad \text{provided} \quad u(x, 0) \equiv 0. \tag{3}$$

The integral is called the convolution of $u(\,\cdot\,, t)$ with $g(t)$ and is usually denoted by $u(x, t) * g(t)$. Clearly

$$v_{xx} = \int_0^t u_{xx}(x, t - y) g(y)\, dy,$$

$$v_t = \int_0^t u_t(x, t - y) g(y)\, dy + u(x, 0) g(t),$$

so that $v_{xx} = v_t$ under the assumptions. An important example is

$$v(x, t) = \int_0^t h(x, t - y) g(y)\, dy.$$

We shall show later that this solution has the property that

$$v(0+, t) = g(t), \qquad v(x, 0+) = 0$$

under suitable continuity restrictions on g.

J The Appell transformation

$$u(x, t) \to v(x, t) = k(x, t) u\left(\frac{x}{t}, -\frac{1}{t}\right), \qquad t > 0.$$

Here $k(x, t)$ is the source solution. We have

$$v_{xx} = \frac{k}{t^2} u_{xx} + \frac{2}{t} u_x k_x + u k_{xx},$$

$$v_t = u k_t + (u_t - x u_x) k t^{-2}.$$

Since $u_{xx} = u_t$ and $k_{xx} = k_t$ we will have $v_{xx} = v_t$ if

$$-\frac{x}{t^2} u_x k = \frac{2}{t} u_x k_x.$$

But this follows from 5(4).

As examples we may transform the heat polynomials v_0 and v_1:

$$v_0 = 1 \to k(x, t), \qquad t > 0;$$

$$v_1 = x \to h(x, t), \qquad t > 0.$$

This transformation of Appell [1892] plays a role in the present theory analogous to that played by inversion

$$u(x, y) \to u\left(\frac{x}{x^2 + y^2}, \frac{y}{x^2 + y^2} \right)$$

in potential theory.

7 DEFINITIONS AND NOTATIONS

We conclude this chapter by introducing several formal definitions. Denote by S an arbitrary region of the x, t-plane.

Definition 7.1

$$u(x, t) \in H \quad \text{in } S$$

$\Leftrightarrow$ A. $u(x, t) \in C^1 \quad \text{in } S,$

         B. $u_{xx} = u_t \quad \text{in } S.$ (1)

If, for example, S is the rectangle R of Theorem 4, then $u(x, t)$ has continuous partial derivatives of order 1 (even on the boundary) and equation (1) holds there. Note that we do not assume the continuity of u_{xt} and u_{tt}. But of course u_{xx} must be assumed to exist for B to have meaning

7. DEFINITIONS AND NOTATIONS

and must then be continuous by A and B. We shall call a function of class H a *temperature function*.

Definition 7.2

$$u(x, t) \in H^* \quad \text{in } S$$

$\Leftrightarrow$ A. $u(x, t) \in C^1$ in S,

          B. $u_{xx} + u_t = 0$ in S.

That is, functions in class H^* satisfy the adjoint heat equation. Obviously

$$u(x, t) \in H \Rightarrow u(x, -t) \in H^*.$$

Let us now complete the definitions of the important functions $k(x, t)$ and $h(x, t)$. In §5 we defined them for $t > 0$ only. Since $k(x, 0+) = 0$ for each fixed $x \neq 0$ and $h(x, 0+) = 0$ for every x, it seems reasonable to define these functions as 0 on the x-axis. We do so, and indeed define them for all negative t as 0. This is a matter of convenience in later work.

Definition 7.3

$$k(x, t) = \frac{e^{-x^2/4t}}{\sqrt{4\pi t}}, \quad t > 0,$$

$$= 0, \quad t \leq 0;$$

$$h(x, t) = \frac{x}{t} k(x, t), \quad -\infty < t < \infty.$$

We show now that these functions, as redefined, are continuous and belong to H over the whole x, t-plane except at $(0, 0)$. We use the following trivial inequality.

Lemma 7

     1. $c > 0, \quad p > 0$

$\Rightarrow$
$$e^{-cy} y^p \leq \left(\frac{p}{ec}\right)^p, \quad 0 \leq y < \infty. \tag{2}$$

This follows since the function on the left of the inequality has a maximum at $y = p/c$.

Theorem 7

$$k(x, t) \in H, \quad (x, t) \neq (0, 0);$$

$$h(x, t) \in H, \quad (x, t) \neq (0, 0).$$

It is clear from the explicit formulas defining these functions that they are of class C^∞ and belong to H at all points off the x-axis. We need only show that

$$\lim k''(x, t) = 0 \tag{3}$$

as $(x, t) \to (x_0, 0)$, $x_0 \neq 0$, in the two-dimensional way. (We make the convention that primes and superscripts, as here used, indicate differentiation with respect to the *space variable* x.) Note that $k''(x_0, 0) = 0$ by Definition 6.3. Given ϵ, we determine δ so that

$$|x - x_0| < \delta, \quad |t| < \delta \Rightarrow |k''(x, t)| < \epsilon. \tag{4}$$

First choose $\delta < |x_0|/2$ so that $|x - x_0| < \delta \Rightarrow |x_0|/2 < |x| < 3|x_0|/2$. From 5(4)

$$k'' = \frac{k}{4t}(x^2 - 2t) = \frac{k}{t^2}(x^2 - 2t)\frac{t}{4}, \quad t > 0,$$

$$|k''| \leq \frac{k}{t^2}\left(\frac{9}{4}x_0^2 + 2t\right)\frac{t}{4}.$$

Now apply Lemma 7 to obtain the maximum of k/t^2, taking $p = 5/2$, $c = x^2/4$, $y = 1/t$:

$$\frac{k(x, t)}{t^2} \leq \frac{1}{\sqrt{4\pi}}\left(\frac{10}{ex^2}\right)^{5/2} = \frac{A}{|x|^5}, \quad t > 0.$$

Hence,

$$|k''(x, t)| \leq \frac{A}{(x_0/2)^5}\left(\frac{9}{4}|x_0|^2 + 2t\right)\frac{t}{4}, \quad t > 0. \tag{5}$$

The factor t now enables us to choose δ to make the right-hand side $< \epsilon$ when $t < \delta$. Since (5) holds trivially when $t \leq 0$, (4) is established. Since $k_{xx} = k_t$ for $t \neq 0$, we see by the continuity just proved that it must also hold for $t = 0$, and $k \in H$ except at the origin.

A similar argument shows that $h(x, t) \in H$ except at the origin. In this case, we use (2) with $p = \frac{7}{2}$. We point out that $h(x, t) \to 0$ as $t \to 0$ even for $x = 0$ since $h(0, t) \equiv 0$. But $h(x, t)$ has a singularity at the origin. Note that it tends to $+\infty$ as $(x, t) \to (0, 0)$ along the parabola $x^2 = t$:

$$\lim_{x \to 0} h(x, x^2) = +\infty.$$

Chapter II

BOUNDARY-VALUE PROBLEMS

1 INTRODUCTION

A boundary-value problem is one in which a function is sought which satisfies a given differential equation in the interior of a region and which satisfies certain prescribed conditions on the boundary of the region. In this chapter we discuss a few of these problems as they apply to the heat equation. A typical region is the rectangle

$$R = \{(x, t) \mid a \leqslant x \leqslant b, \ 0 \leqslant t \leqslant c\}.$$

At first sight we might hope to find a temperature function taking prescribed values on the boundary of R. But a moment's thought about the physical meaning of the problem shows that this is too much to hope. This would be to require that a bar with prescribed temperatures initially, $t = 0$, and with its ends at $x = a$ and $x = b$ held at given temperatures, would end up with any temperature we liked! From physical considerations we expect rather that the temperature at any time is determined by the initial temperature and by what is done to the two ends of the rod. Accordingly, we formulate the following more sensible problem, usually called the *first boundary-value problem* for the heat equation.

Problem I Find $u(x, t) \in H$, (x, t) interior to R, such that

A. $u(x, 0+) = \varphi(x)$, $a < x < b$;
B. $u(a+, t) = g_1(t)$, $0 < t < c$;
C. $u(b-, t) = g_2(t)$, $0 < t < c$.

Here the functions $f(x)$, $g_1(t)$, $g_2(t)$ are prescribed. In spite of our intuition, it will develop that Problem I does not always have a unique solution. We shall discuss in the next section how to modify it to guarantee uniqueness. Note that condition A might more properly be referred to as an initial condition since it refers to time rather than to space. Conditions B and C might be replaced by others prescribing the "flux" $\partial u/\partial x$ at $x = a$ and $x = b$. This is called the second boundary-value problem. A third arises if the "radiation" of heat at the ends of the bar is assumed known. For then, according to Newton's Law of radiation (flux proportional to temperature difference between the bar and the surrounding medium), a linear combination of u and u_x, $u + \lambda u_x$, would be prescribed at the ends of the bar. We do not set these problems down explicitly, for we shall concentrate mainly on Problem I.

There are important limiting cases of Problem I. If the bar is very long the effect of the end conditions on the central portion becomes insignificant, and as a limiting situation we may assume the bar as covering the entire x-axis from $-\infty$ to $+\infty$, so that conditions B and C disappear. We thus revert to the Cauchy problem on a characteristic, as described in Chapter I, §4:

$$u(x, t) \in H, \quad 0 < t < c;$$
$$u(x, 0+) = \varphi(x), \quad -\infty < x < \infty.$$

Finally, the bar may be semi-infinite, on the positive x-axis, say. The problem becomes

$$u(x, t) \in H, \quad 0 < x < \infty, \quad 0 < t < c;$$
$$u(x, 0+) = \varphi(x), \quad 0 < x < \infty,$$
$$u(0+, t) = g(t), \quad 0 < t < c.$$

2 UNIQUENESS

To see that a solution of Problem I need not be unique consider the function 5(6) of Chapter I:

$$\varphi(x, t) = \sum_{n=-\infty}^{\infty} h(x + 2n\pi, t) \tag{1}$$

on the rectangle R with $a = 0$, $b = 2\pi$. Since $h(x, 0 +) = 0$, one may expect that $\varphi(x, 0 +) = 0$ (if the series converges uniformly on $0 \leq t \leq \delta$). Moreover, if we rewrite series (1) as

$$\varphi(x, t) = h(x, t) + \sum_{n=1}^{\infty} [h(x + 2n\pi, t) + h(x - 2n\pi, t)]$$

we see that $\varphi(0, t) = 0$ since $h(-x, t) = -h(x, t)$. We observed earlier that $\varphi(x, t)$ is periodic in the variable x, so that $\varphi(2\pi, t) = 0$. Thus φ vanishes on three sides of R and is not identically zero. Consequently, it can be added to any solution of Problem I to obtain another. The assumptions here made about series (1) will be proved in Chapter V.

The fact cited earlier that $\varphi(0, 0 +) = 0$ but that $\varphi(x, x^2) \to \infty$ as $x \to 0 +$ should suggest that an alteration of Problem I might be in order. Temperature functions may behave well on the boundary of R when the approach thereto is along certain curves interior to R, but badly when the approach is along others. Accordingly, we alter Problem I to demand continuity on the boundary of R.

Problem I

$$u(x, t) \in H, \quad (x, t) \text{ interior to } R;$$
$$u(x, t) \in C, \quad (x, t) \text{ throughout } R;$$

A. $u(x, 0) = \varphi(x), \quad a \leq x \leq b;$
B. $u(a, t) = g_1(t), \quad 0 \leq t \leq c;$
C. $u(b, t) = g_2(t), \quad 0 \leq t \leq c.$

If there is to be a solution of this altered problem it is clear that the prescribed functions cannot be completely arbitrary. They must be such that $u \in C$ on the boundary of R. Thus $\varphi(a +) = g_1(0 +)$ and $\varphi(b -) = g_2(0 +)$. Henceforth, it is this altered version which we shall call the first boundary-value problem for the finite bar. We shall show later that any solution thereof is unique. Using Theorem 4 of Chapter I, we prove at once a slightly weaker result. Green's theorem as proved there required continuity of u_{xx} on the boundary of R.

Theorem 2

1. $u(x, t) \in H, \quad a \leq x \leq b, \ 0 \leq t \leq c,$
2. $u(a, t) = u(b, t) = 0, \quad 0 \leq t \leq c,$
3. $u(x, 0) = 0, \quad a \leq x \leq b,$

$\Rightarrow \qquad u(x, t) = 0, \quad a \leq x \leq b, \ 0 \leq t \leq c.$

For an arbitrary $t_0 < c$ apply Green's theorem to the rectangle

$$R = \{(x, t) \mid a \leqslant x \leqslant b, \ 0 \leqslant t \leqslant t_0\}, \qquad 0 < t_0 < c,$$

replacing u by u^2 and v by 1. Then

$$L[u^2] = 2u_x^2 + 2u[u_{xx} - u_t] = 2u_x^2,$$

$$L^*[1] = 0,$$

$$\iint_R u_x^2 \, dx \, dt = \int_\Gamma u^2 \, dx + 2uu_t \, dt,$$

$$\iint_R u_x^2 \, dx \, dt = -\int_a^b u^2(x, t_0) \, dx. \qquad (2)$$

Since $u = 0$ on three sides of R, the line integral over Γ reduces to the single integral (2). Since both integrands are $\geqslant 0$ and since $u(x, t_0) \in C$ on $a \leqslant x \leqslant b$, both integrals (2) are zero and $u(x, t_0) \equiv 0$, $a \leqslant x \leqslant b$. Since t_0 was arbitrary, this completes the proof. To prove the uniqueness theorem in full generality, we take a different approach in the following section.

3 THE MAXIMUM PRINCIPLE

We show here that if $u(x, t) \in H$ interior to the rectangle R and $\in C$ on its boundary Γ, then it takes on its maximum and minimum values either on the vertical sides or on the base of R. This will follow as a consequence of the following result.

Theorem 3.1

1. $u(x, t) \in H$, $\qquad\qquad\qquad a < x < b, \ 0 < t < c,$
2. $u(x, t) \in C$, $\qquad\qquad\qquad a \leqslant x \leqslant b, \ 0 \leqslant t \leqslant c,$
3. $u(a, t) \geqslant 0, \ u(b, t) \geqslant 0, \quad 0 \leqslant t \leqslant c,$
4. $u(x, 0) \geqslant 0,$ $\qquad\qquad\qquad a \leqslant x \leqslant b,$

$\Rightarrow \qquad u(x, t) \geqslant 0, \qquad a \leqslant x \leqslant b, \ 0 \leqslant t \leqslant c. \qquad (1)$

Since hypothesis 1 refers only to the interior of R, there are no differentiability assumptions about u on the line $t = c$. However, if we can prove (1) for interior points of R, it will also hold for $t = c$ by hypothesis 2. Hence, it is no restriction to assume $u \in H$ even on the line $t = c$. We shall need this in the proof.

For an arbitrary $\epsilon > 0$ define

$$v(x, t) = u(x, t) + \epsilon t.$$

This continuous function has a minimum value in the closed rectangle R, taken on say, at (x_1, t_1). We show that $x_1 = a$ or b, or else $t_1 = 0$. Suppose the contrary: $a < x_1 < b$, $0 < t_1 \leq c$. At such an interior minimum $v_{xx} \geq 0$ and $v_t \leq 0$ by elementary calculus ($v_t = 0$ if $t_1 < c$). But by 1 (as modified if $t_1 = c$)

$$v_{xx}(x_1, t_1) - v_t(x_1, t_1) = L[u] - \epsilon = -\epsilon.$$

This is impossible since the left-hand side is ≥ 0. Hence (x_1, t_1) must be a point where hypothesis 3 or 4 applies. Consequently, if (x, t) is any point of R,

$$u(x, t) = v(x, t) - \epsilon t \geq v(x_1, t_1) - \epsilon t$$

$$= u(x_1, t_1) + \epsilon t_1 - \epsilon t \geq \epsilon(t_1 - t).$$

Since ϵ was arbitrary, this inequality implies that $u(x, t) \geq 0$, as we wished to prove. If the inequalities are reversed in hypotheses 3 and 4 ($u \leq 0$), they will be reversed in the conclusion. For, we may apply the theorem to $-u(x, t)$ which also belongs to H.

Let us denote the vertical sides plus the base of the rectangle R by γ (the total boundary being Γ). Then an immediate consequence of Theorem 6 is the following. If $u_1(x, t)$ and $u_2(x, t)$ satisfy hypotheses 1 and 2, then

$$u_1(x, t) \leq u_2(x, t), \quad (x, t) \in \gamma \Rightarrow u_1(x, t) \leq u_2(x, t), \quad (x, t) \in R. \tag{2}$$

Since constants are temperature functions, we have as special cases

$$m \leq u(x, t) \leq M, \quad (x, t) \in \gamma \Rightarrow m \leq u(x, t) \leq M, \quad (x, t) \in R. \tag{3}$$

This is the analytic expression of the so-called *maximum* (*minimum*) *principle* for temperature functions, as here applied to a rectangle. If M is the maximum value of $u(x, t)$ for $(x, t) \in \gamma$, then it is also the maximum value for $(x, t) \in R$. Roughly stated, the maximum value of a temperature function over a region is assumed on the boundary thereof.

It is useful to modify hypothesis 2 to permit use in some cases when $u(x, t)$ approaches no limit as (x, t) approaches boundary points of R. Such a case is $k(x, t)$ which has no limit as $(x, t) \to (0, 0)$ in the two-dimensional way. Yet $\underline{\lim}\, k(x, t) \geq 0$, so that the following modification of the theorem is suggested.

Corollary 3.1

1. $u(x, t) \in H$ inside R,

2. $\lim\limits_{(x, t) \to (x_0, t_0)} u(x, t) \geq 0$ for every (x_0, t_0) on γ,

$\Rightarrow$ $u(x, t) \geq 0, \quad (x, t) \in R.$

From the definition of limit inferior, if $\epsilon > 0$ then for each (x_0, t_0) on γ there is a δ_0 such that $u(x, t) \geq -\epsilon$ for all (x, t) of R within a distance δ_0 of (x_0, t_0). By the Heine–Borel theorem, there is a single δ applicable to all points of γ. Hence, for a rectangle $R^* \subset R$, we have $u(x, t) \geq -\epsilon$ on γ^*. Applying Theorem 3.1 to the function $u(x, t) + \epsilon$ on R^*, we conclude that

$$u(x, t) \geq -\epsilon \quad \text{in } R^*.$$

Since ϵ is arbitrary, this gives the desired result.

As an immediate consequence of these considerations we can now establish uniqueness for Problem I, as restated in §2.

Theorem 3.2

1. $u(x, t) \in H$, (x, t) interior to R,
2. $u(x, t) \in C$, (x, t) throughout R,
3. $u(x, 0) = 0$, $a \leq x \leq b$,
4. $u(a, t) = u(b, t) = 0$ $0 \leq t \leq c$,

$\Rightarrow$ $u(x, t) \equiv 0, \quad (x, t) \in R.$

This follows from (3) with $m = M = 0$.

As a further consequence of Theorem 3.1, we can now draw the important conclusion that a solution of the first boundary-value problem depends continuously on its boundary and initial conditions. That is, a small change in the prescribed functions $\varphi(x)$, $g_1(t)$, $g_2(t)$ of Problem I will produce a small change in the solution $u(x, t)$. More precisely, if u_1 and u_2 both satisfy hypotheses 1 and 2 of Theorem 3.1 and if

$$|u_1(x, t) - u_2(x, t)| \leq \epsilon, \qquad (x, t) \in \gamma, \tag{4}$$

then the same inequality must hold for all $(x, t) \in R$. This follows from (3) since $\epsilon \in H$ and inequality (4) is equivalent to

$$u_2(x, t) - \epsilon \leq u_1(x, t) \leq u_2(x, t) + \epsilon.$$

4 A CRITERION FOR TEMPERATURE FUNCTIONS

If in Green's formula [4(4) of Chapter I] we take $u \in H$ and $v = 1$, the double integral vanishes and the line integral becomes

$$\int_\Gamma u\, dx + u_x\, dt = 0.$$

This equation may be used to characterize temperature functions u. It has the advantage over the definition of H that it does not postulate the continuity of u_t nor the existence of u_{xx}. As usual, we define Γ as the boundary of the rectangle

$$R = \{(x, t) \mid a \leq x \leq b, \ c \leq x \leq d\}.$$

Theorem 4

1. $u(x, t), u_x(x, t) \in C$ in a domain D,
2. $\int_\Gamma u\, dx + u_x\, dt = 0$ every R in D,

$\Rightarrow \qquad\qquad u(x, t) \in H \qquad$ in D.

Let (x_1, t_1) be an arbitrary point of D. Include it in a rectangle R lying entirely in D and set

$$U(x, t) = \int_{a, c}^{x, t} u\, dx + u_x\, dt,$$

where integration is along any broken line with segments parallel to the axes (Figure 2). By hypothesis 2, $U(x, t)$ is single-valued, and by elementary calculus [Widder, 1961c; 228],

$$U_x = u, \qquad U_t = u_x.$$

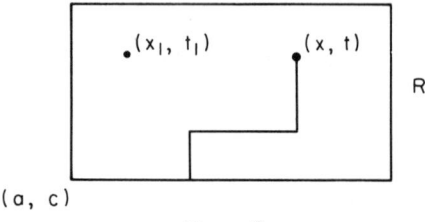

Figure 2

By 1 we see that $U_{xx} = u_x \in C$ and hence $U \in H$ in R. We shall see in Chapter IV, §13 that all derivatives of a temperature function exist and are continuous, so that

$$u_{xx} = U_{tx}, \qquad u_t = U_{xt},$$

and $u \in H$ in R and hence at the arbitrary point (x_1, t_1). This completes the proof.

5 SOLUTION OF PROBLEM I IN A SPECIAL CASE

We give here a complete solution of the first boundary-value problem as described in §2, when $g_1(t) = g_2(t) = 0$ and $\varphi(x) \in C^2$. Physically, this is the problem of determining the subsequent temperatures of a finite bar with two ends immersed in ice, the initial temperature of the bar being known.

Theorem 5

1. $u(x, t) \in H$, $\qquad 0 < x < \pi, \quad 0 < t < c,$
2. $u(x, t) \in C$, $\qquad 0 \leq x \leq \pi, \quad 0 \leq t \leq c,$
3. $u(0, t) = u(\pi, t) = 0$, $\qquad 0 \leq t \leq c,$
4. $u(x, 0) = \varphi(x) \in C^2$, $\qquad 0 \leq x \leq \pi,$

$$\Leftrightarrow \quad u(x, t) = \sum_{n=1}^{\infty} a_n e^{-n^2 t} \sin nx, \qquad 0 \leq x \leq \pi, \quad 0 \leq t \leq c, \qquad (1)$$

$$a_n = \frac{2}{\pi} \int_0^{\pi} \varphi(y) \sin ny \, dy, \qquad n = 1, 2, \ldots, \qquad (2)$$

where $\varphi(y) \in C^2$, $0 \leq y \leq \pi$, and $\varphi(0) = \varphi(\pi) = 0$.

Note that if conditions 2–4 are to hold simultaneously, then $\varphi(x)$ must vanish at 0 and π. We assume first the representations (1) and (2), and

5. SOLUTION OF PROBLEM I IN A SPECIAL CASE

prove the necessity of the conditions. The sufficiency will then follow by use of Theorem 3.2 which states that there can be at most one solution to Problem I.

It is a familiar fact that the Fourier coefficients (2) remain bounded when multiplied by n^2:

$$a_n = O\left(\frac{1}{n^2}\right), \qquad n \to \infty.$$

This follows, under our assumptions about φ, by integration by parts:

$$a_n = -\frac{2}{n^2\pi} \int_0^\pi \varphi''(y) \sin ny \, dy,$$

$$|a_n| \leq \frac{M}{n^2}, \qquad M = 2 \max_{0 \leq y \leq \pi} |\varphi''(y)|.$$

Hence

$$u(x, t) = \sum_{n=1}^\infty a_n e^{-n^2 t} \sin nx \ll M \sum_{n=1}^\infty \frac{1}{n^2},$$

$$-\infty < x < \infty, \quad 0 \leq t < \infty,$$

and series (1) converges uniformly over the half-plane $t \geq 0$. Hence condition 2 is immediate. By inspection, $u(0, t) = u(\pi, t) = 0$ for all $t \geq 0$. Also, condition 4,

$$u(x, 0) = \sum_{n=1}^\infty a_n \sin nx = \varphi(x), \qquad 0 \leq x \leq \pi, \tag{3}$$

follows by the theory of Fourier series [Widder, 1961c; 406]. Differentiating series (1), we have for an arbitrary $\delta > 0$

$$u_{xx} = u_t = -\sum_{n=1}^\infty a_n n^2 e^{-n^2 t} \sin nx \ll M \sum_{n=1}^\infty e^{-n^2 \delta}, \qquad \delta \leq t < \infty. \tag{4}$$

The dominant series clearly converges ($\lim_{n \to \infty} n^2 e^{-n^2 \delta} = 0$) and is independent of x and t. Hence the differentiated series (4) converges uniformly in the half-plane $t \geq \delta$, so that the term-by-term differentiation is valid. Since δ was arbitrary, $u(x, t) \in H$ for $t > 0$ and condition 1 is satisfied (liberally!).

Observe that hypothesis 4 could be considerably weakened. For example, if $\varphi(x)$ is continuous and of bounded variation on $0 \leq x \leq \pi$ with

$\varphi(0) = \varphi(\pi) = 0$, then the Fourier series (3) converges to $\varphi(x)$ on $0 \leq x \leq \pi$ by Jordan's theorem and $u(x, 0 +) = \varphi(x)$ there by Abel's theorem. We do not give details here, for we shall be able to show later that $u(x, 0 +) = \varphi(x)$ even in some cases when series (3) diverges. For the present, we can already draw the remarkable conclusion that for each fixed $t > 0$ the function (1) can be extended analytically into the complex plane as a function of x and is entire.

Corollary 5

1. $u(x, t)$ satisfies conditions 1–4 of Theorem 5,
2. $s = \sigma + i\tau$,
3. $0 < t_0 < c$,

$\Rightarrow \qquad u(s, t_0) \in A \quad$ (analytic), $\quad |s| < \infty.$

If $|s| \leq R$, then

$$|\sin ns| = \frac{|e^{ni(\sigma + i\tau)} - e^{-ni(\sigma + i\tau)}|}{2} \leq e^{n|\tau|} \leq e^{nR}.$$

Hence

$$\sum_{n=1}^{\infty} a_n e^{-n^2 t_0} \sin ns \ll M \sum_{n=1}^{\infty} n^{-2} e^{-n^2 t_0 + nR}, \qquad |s| \leq R. \tag{5}$$

The dominant series of constants coverges, so that the series of functions (5) converges uniformly in the disk $|s| \leq R$. Its sum is therefore analytic for $|s| < R$ by Weierstrass's theorem. Since R is arbitrary, we conclude that $u(s, t_0)$ is entire, as stated.

We shall see in Chapter III, §16 that not all temperature functions $u(x, t)$ can be extended analytically so as to be entire, as in the present case. The essential hypothesis here is that u vanishes on the vertical sides of the given rectangle. This enables its definition to be extended, as a function of H, outside the given rectangle, using the *principle of reflection* (to be discussed later). The series (1) clearly converges in the whole half-plane $t > 0$ and may be used to define u there as a function of H.

6 UNIQUENESS FOR THE INFINITE ROD

We discuss here the uniqueness question for Problem I in the limiting case when $a = -\infty$, $b = +\infty$ or when $a = 0$, $b = +\infty$. As pointed out

6. UNIQUENESS FOR THE INFINITE ROD

in §1, the first case is equivalent to the Cauchy problem on a characteristic:

$$u(x, t) \in H, \quad t > 0;$$

$$u(x, 0+) = \varphi(x), \quad -\infty < x < \infty.$$

That a solution of this problem need not be unique is already evident since $h(x, 0+) = 0$ for all x. Accordingly, we alter the problem as in §2 for the finite bar, by demanding in addition that $u(x, t)$ should be continuous for $t \geq 0$. But this additional requirement is still insufficient to produce uniqueness. There exist so-called *null solutions* of the heat equation, not identically zero but vanishing over an entire characteristic. Here is one defined by a definite integral, constructed by Rosenbloom and Widder [1958; 607]:

$$u(x, t) = \int_0^\infty g(xy, ty^2) a(y) \, dy,$$

$$g(x, t) = e^x \cos(x + 2t) + e^{-x} \cos(x - 2t),$$

$$a(y) = \exp(-y^{4/3}) \cos(\sqrt{3} \, y^{4/3}).$$

Here $u(x, t) \in H$ for all x and t and $u(x, 0) = 0$ for $-\infty < x < \infty$. We shall not prove these facts here. Another example where facts are more transparent is

$$u(x, t) = \sum_{n=0}^\infty f^{(n)}(t) \frac{x^{2n}}{(2n)!};$$

$$f(t) = e^{-1/t^2}, \quad t \neq 0;$$

$$f(0) = 0.$$

If term-by-term differentiation is valid

$$u_{xx} = u_t = \sum_{n=0}^\infty f^{(n+1)}(t) \frac{x^{2n}}{(2n)!}, \quad -\infty < t < \infty. \tag{1}$$

Since $f(t)$ is known to vanish with all its derivatives at $t = 0$, obviously $u(x, 0) = 0$. Evidently, $u(x, t)$ is not identically zero. In fact, $u(0, t) = f(t) > 0$, $t \neq 0$. The validity of equation (1) will be established in Chapter III, Theorems 11.2 and 14.

This example makes it clear that no mere continuity requirements on u or its derivatives will suffice to produce uniqueness in the present case. We prove now that a condition limiting the rate of growth of $u(x, t)$ as $|x| \to \infty$ will be effective.

Theorem 6.1

1. $u(x, t) \in H$, $0 < t < c$;
2. $u(x, t) \in C$, $0 \leq t < c$;
3. $|u(x, t)| < Me^{ax^2}$,   some M, a, $-\infty < x < \infty$, $0 \leq t < c$;
4. $u(x, 0) = 0$, $-\infty < x < \infty$,

$\Rightarrow$ $u(x, t) \equiv 0$, $-\infty < x < \infty$, $0 \leq t < c$.

Choose a constant $A > a$, $A > 1/(4c)$, and form the function

$$v(x, t) = \frac{e^{Ax^2/(1-4At)}}{\sqrt{1-4At}} = \sqrt{4\pi}\, k(\lambda x, 1 + \lambda^2 t), \quad \lambda = i\sqrt{4A}.$$

Appealing to transformations E and F of Chapter I, §5, we see that $v(x, t) \in H$ for $0 \leq t < 1/(4A)$. Since $v(x_0, t) \in \uparrow$, we have

$$v(x, t) \geq v(x, 0) = e^{Ax^2}, \quad 0 \leq t < \frac{1}{4A}. \tag{2}$$

Let (x_0, t_0) be an arbitrary point in the strip $0 < t < 1/(4A)$ and choose $L > |x_0|$. Compare the functions $|u(x, t)|$ and

$$w(x, t) = Me^{(a-A)L^2} v(x, t)$$

on the contour γ (sides and base) of the rectangle

$$R = \{(x, t) \mid -L \leq x \leq L, \ 0 \leq t \leq 1/(4A)\}.$$

On the base

$$|u(x, 0)| = 0 < w(x, 0) = Me^{(a-A)L^2} e^{Ax^2}.$$

On the vertical sides, by 3,

$$|u(\pm L, t)| < Me^{aL^2}.$$

Also by (2)

$$w(\pm L, t) = Me^{(a-A)L^2} v(\pm L, t) \geq Me^{aL^2}.$$

That is,

$$-w(x, t) < u(x, t) < w(x, t) \qquad (3)$$

on γ and hence inside R by 3(2). In particular, (3) holds at (x_0, t_0), so that

$$|u(x_0, t_0)| < Me^{(a-A)L^2} v(x_0, t_0) = o(1), \qquad L \to \infty. \qquad (4)$$

If we keep (x_0, t_0) fixed and allow L to become infinite, (4) implies that $u(x_0, t_0) = 0$. Since (x_0, t_0) was an arbitrary point, we have proved $u(x, t) \equiv 0$ in the strip $0 < t < 1/(4A)$. Now repeat the above argument as applied to the function $u(x, t + t_0)$, where $t_0 < 1/(4A)$. The conclusion will be that $u(x, t) \equiv 0$ for $0 < t < 2/(4A)$. Repeating the argument n times if necessary, we prove $u(x, t) \equiv 0$ for $0 < t < n/(4A)$. Clearly any point of the given strip, $0 < t < c$, can be included in one of these strips, so the proof is complete.

As an immediate consequence of Theorem 6.1, we obtain a uniqueness theorem for the semi-infinite rod.

Theorem 6.2

1. $u(x, t) \in H,\qquad 0 < t < c,\ 0 < x < \infty,$
2. $u(x, t) \in C,\qquad 0 \leqslant t < c,\ 0 \leqslant x < \infty,$
3. $|u(x, t)| < Me^{ax^2},\quad$ some $M, a,\ 0 \leqslant t < c,\ 0 \leqslant x < \infty,$
4. $u(0, t) = 0,\qquad 0 \leqslant t < c,$
5. $u(x, 0) = 0,\qquad 0 \leqslant x < \infty,$

$\Rightarrow \qquad u(x, t) \equiv 0, \qquad 0 \leqslant t < c,\ 0 \leqslant x < \infty.$

Define $u(x, t)$ for negative x so that u becomes an odd function of x:

$$u(-x, t) = -u(x, t).$$

The extended function $\in H$ in the strip $0 < t < c$ except perhaps on the t-axis. We shall see in Chapter VI, §7, when the "reflection principle" has been discussed, that $u \in H$ even on the t-axis, by virtue of hypothesis 4. We can now apply Theorem 6.1 to the extended function to conclude that it vanishes in the strip $0 \leqslant t < c$. The original function vanishes in the half-strip, as stated.

Chapter **III**

FURTHER DEVELOPMENTS

1 INTRODUCTION

We have already seen the importance of the source solution $k(x, t)$ and of the derived source solution $h(x, t)$. In this chapter, we study their properties in detail and compute their Laplace transforms. After giving a brief introduction to operational calculus, we use these transforms to conjecture series and integral solutions of the heat equation. We give conditions for the validity of the series expansions, postponing to a later chapter the detailed study of the integrals.

2 THE SOURCE SOLUTION

Let us collect here the essential properties of the source solution,

$$k(x, t) = \frac{e^{-x^2/4t}}{\sqrt{4\pi t}}, \quad t > 0,$$

$$= 0, \quad t \leq 0.$$

2. THE SOURCE SOLUTION

In Theorem 7 of Chapter I, we showed that it is a temperature function for all (x, t) except $(0, 0)$. Other properties are immediate as follows:

Theorem 2

A. $k(x, t) > 0,$ $\qquad\qquad t > 0;$

B. $\lim_{t \to 0} k(x, t) = 0,$ $\qquad x \neq 0;$

C. $\lim_{t \to 0+} k(0, t) = \infty,$

D. $\int_{-\infty}^{\infty} k(x, t) \, dx = 1,$ $\qquad t > 0;$

E. $\lim_{t \to 0+} \int_{-\delta}^{\delta} k(x, t) \, dx = 1,$ $\qquad \delta > 0.$

Conclusions A–C follow by inspection of the definition. Change of variable gives

$$\int_{-\infty}^{\infty} k(x, t) \, dx = \frac{1}{\sqrt{\pi}} \int_{-\infty}^{\infty} e^{-r^2} \, dr, \tag{1}$$

$$\int_{-\delta}^{\delta} k(x, t) \, dx = \frac{1}{\sqrt{\pi}} \int_{-\delta/\sqrt{4t}}^{\delta/\sqrt{4t}} e^{-r^2} \, dr.$$

The value of the probability integral (1) is $\sqrt{\pi}$ so that D and E are proved.

These facts about $k(x, t)$ enable us to interpret its physical significance. If $k(x, t)$ is the temperature of an infinite bar at a point x and at time t, then by 3(5) of Chapter I, the total quantity of heat in the bar at time t is

$$c\rho \int_{-\infty}^{\infty} k(x, t) \, dx = c\rho.$$

Moreover, B and E show that for small t most of the heat is concentrated near $x = 0$. If $c\rho = 1$, we may thus describe $k(x, t)$ as the temperature of the bar, initially at temperature zero, produced by introducing unit quantity of heat into the bar at $x = 0$. The total quantity of heat in the bar remains 1. The temperature at the origin is instantaneously enormous (by C) but levels off rapidly. The graph of $k(x, t)$ as a function of x is that of the familiar probability distribution (Figure 3). The reason for the term *source-solution* is now evident. There is a *source* at $x = 0$. We say that the strength of the source is 1 on account of E.

It is noteworthy by A that the effect of introducing a quantity of heat at $x = 0$ is instantaneously noticeable at remote points of the bar; that is,

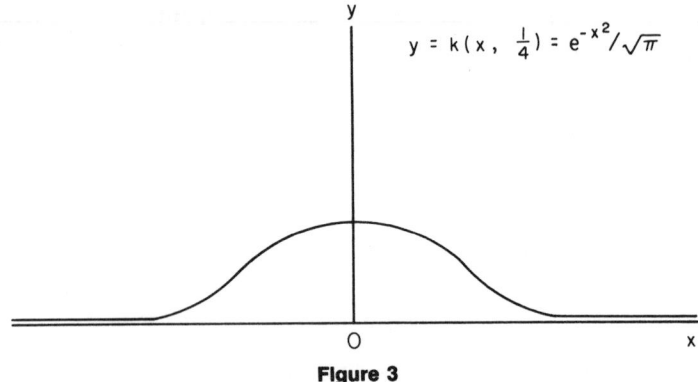

Figure 3

heat from a source is propogated at infinite velocity. This paradox serves to emphasize our remark in Chapter I, §2, that Postulates A and B are only approximations to physical conditions. The theory is nonetheless useful in practice. In the present instance, the temperature $k(x, t)$ falls off very rapidly as $|x|$ increases and is negligible for large $|x|$ and small t.

3 THE ADDITION FORMULA FOR $k(x,t)$

A very important integral for the study of heat conduction is

$$u(x, t) = \int_{-\infty}^{\infty} k(x - y, t)\varphi(y)\, dy. \tag{1}$$

This has the form of a *bilateral convolution* between the functions $k(x, t)$ and $\varphi(x)$, the convolution bearing on the space variable x. The integral (1) is usually abbreviated by the symbol $k(x, t) * \varphi(x)$. Compare 6(2) of Chapter I where the convolution there introduced was with respect to the time variable and was *unilateral*. The ambiguity in notation usually causes no confusion.

Equation (1) also defines the *Poisson transform*, which will be studied carefully in Chapter IV. Here we may already see that it provides a likely solution of the Cauchy problem with initial condition $u(x, 0) = \varphi(x)$. If we disregard niceties, $u(x, t)$ should belong to H by Method C (Chapter I, §6). Moreover,

$$k(x, t) * \varphi(x) = \varphi(x) * k(x, t) = \int_{-\infty}^{\infty} k(y, t)\varphi(x - y)\, dy$$

$$= \frac{1}{\sqrt{\pi}} \int_{-\infty}^{\infty} e^{-r^2}\varphi(x - r\sqrt{4t})\, dr, \qquad \frac{y^2}{4t} = r^2.$$

3. THE ADDITION FORMULA FOR $k(x,t)$

Hence, for suitable functions φ we would expect

$$u(x, 0+) = \frac{1}{\sqrt{\pi}} \int_{-\infty}^{\infty} e^{-r^2} \varphi(x) \, dr = \varphi(x).$$

In particular, if $\varphi(x) = k(x, t_1)$, $t_1 > 0$, the function

$$u(x, t) = \int_{-\infty}^{\infty} k(x - y, t) k(y, t_1) \, dy$$

belongs to H for $t > 0$ and reduces to $k(x, t_1)$ when $t \to 0+$. But we know one such function to be $k(x, t + t_1)$. Could there be another? No, not in the light of Theorem 6.1, Chapter II. Apply that theorem to the difference $u(x, t) - k(x, t + t_1)$ with $M = (4\pi t_1)^{-1/2}$ and $a = 0$. Clearly

$$0 < k(x, t + t_1) \leq \frac{1}{\sqrt{4\pi(t + t_1)}} \leq \frac{1}{\sqrt{4\pi t_1}}, \qquad t \geq 0,$$

$$0 < u(x, t) \leq \int_{-\infty}^{\infty} k(x - y, t) \frac{1}{\sqrt{4\pi t_1}} \, dy = \frac{1}{\sqrt{4\pi t_1}}.$$

The difference is consequently identically zero, and $u(x, t) = k(x, t + t_1)$. We then conjecture the following result:

Theorem 3

1. $t_1 > 0, \qquad t_2 > 0$

$\Rightarrow \qquad k(x, t_1) * k(x, t_2) = k(x, t_1 + t_2).$

We prove the result algebraically. We have

$$k(x, t_1) * k(x, t_2) = \frac{1}{4\pi\sqrt{t_1 t_2}} \int_{-\infty}^{\infty} e^{-(x-y)^2/(4t_1)} e^{-y^2/(4t_2)} \, dy$$

$$= \frac{\exp(4t_1 B^2 - x^2)/(4t_1)}{4\pi\sqrt{t_1 t_2}} \int_{-\infty}^{\infty} e^{-(Ay+B)^2} \, dy, \quad (2)$$

where

$$A^2 = \frac{t_1 + t_2}{4t_1 t_2}, \qquad B^2 = \frac{t_2 x^2}{4t_1(t_1 + t)}.$$

III. FURTHER DEVELOPMENTS

The probability integral (2) has the value $\sqrt{\pi}/A$ so that

$$k(x, t_1) * k(x, t_2) = \frac{\exp[-x^2/4(t_1 + t_2)]}{\sqrt{4\pi(t_1 + t_2)}}$$

$$= k(x, t_1 + t_2).$$

This concludes the proof.

4 THE HOMOGENEITY OF $k(x,t)$

From the definition of $k(x, t)$ it is evident that for every $\lambda > 0$,

$$k(\lambda x, \lambda^2 t) = \frac{k(x, t)}{\lambda}.$$

Let us inquire if $k(x, t)$ is the only solution of the heat equation having this homogeneity property: $k(x, t^2)$ is homogeneous of degree -1. The answer is negative, as proved in the following theorem.

Theorem 4

1. $u(x, t) \in H$, $\qquad t > 0$,
2. $u(\lambda x, \lambda^2 t) = \dfrac{u(x, t)}{\lambda}$, $\quad$ all $\lambda > 0$, $\ t > 0$, $\hfill (1)$

$\Leftrightarrow \quad u(x, t) = k(x, t)\left[A + B\int_0^{x/\sqrt{4t}} e^{r^2} dr\right]$, $\quad A, B$ constants.

We prove the sufficiency of the conditions. Set $\lambda = (4t)^{-1/2}$ in (1),

$$u(x, t) = \frac{f(x/\sqrt{4t})}{\sqrt{4t}}, \qquad f(x) = u(x, \tfrac{1}{4}). \hfill (2)$$

From hypothesis 1, $u_{xx} = u_t$, or

$$f''(z) + 2zf'(z) + 2f(z) = 0, \qquad z = \frac{x}{\sqrt{4t}}. \hfill (3)$$

This is Hermite's differential equation. One solution is e^{-z^2}, the one that yields $k(x, t)$ through (2). By a familiar process it is possible to determine $g(z)$ so that $e^{-z^2}g(z)$ will be a second solution of (3). We find

$$g(z) = \int_0^z e^{r^2} dr,$$

and we may verify by direct substitution that $f(z) = e^{-z^2}g(z)$ does indeed satisfy equation (3). The complete solution of Hermite's equation is therefore

$$f(z) = Ae^{-z^2} + Be^{-z^2}\int_0^z e^{r^2}\,dr,$$

where A and B are any constants. Substitution in (2) gives the desired result. The necessity of the conditions follows by direct verification, or by reversing the above calculations.

Corollary 4

$\Rightarrow$
1. $u(x, t) \in H,$ $\qquad t > 0,$
2. $u(\lambda x, \lambda^2 t) = \dfrac{u(x, t)}{\lambda},$ $\qquad \lambda > 0, \ t > 0,$
3. $u(x, t) > 0,$ $\qquad t > 0,$

$$u(x, t) = Ak(x, t), \quad \text{some } A > 0.$$

We need only show, as a consequence of the additional hypothesis 3, that the constant B of Theorem 4 is zero and that $A > 0$. By 3

$$\frac{u(x, t)}{k(x, t)} > 0, \qquad t > 0, \tag{4}$$

so that from the conclusion of Theorem 4

$$\lim_{x \to +\infty} \frac{u(x, t)}{k(x, t)} = -\infty, \qquad \text{if } B < 0,$$

$$\lim_{x \to -\infty} \frac{u(x, t)}{k(x, t)} = -\infty, \qquad \text{if } B > 0.$$

Since either result contradicts 3, $B = 0$. That $A > 0$ now follows from (4). Thus $k(x, t)$ may be characterized as the only positive temperature function $u(x, t)$ which is homogeneous of degree -1 and for which

$$\int_{-\infty}^{\infty} u(x, t)\,dx = 1, \qquad t > 0.$$

5 AN INTEGRAL REPRESENTATION OF $k(x,t)$

In Chapter I, §6, we gave an integral representation of $k(x, t)$ without proof. Let us establish it here.

Theorem 5

1. $-\infty < x < \infty, \quad 0 < t < \infty$

$$k(x, t) = \frac{1}{\pi} \int_0^\infty e^{-ty^2} \cos xy \, dy \qquad (1)$$

$$= \frac{1}{2\pi} \int_{-\infty}^\infty e^{-ty^2 + xyi} \, dy \qquad (2)$$

$$= \frac{1}{2\pi} \int_0^\infty e^{-ty} \frac{\cos x\sqrt{y}}{\sqrt{y}} \, dy. \qquad $$

That is, $k(x, t)$ as a function of x is the cosine-transform of $e^{-ty^2}/2$; as a function of t it is the Laplace transform of $\cos x\sqrt{y}/(2\pi\sqrt{y})$. We give two proofs.

The integral (1) clearly converges for all x and all $t > 0$. Denote its value by $u(x, t)$ and compute $u_x(x, t)$:

$$u_x(x, t) = -\frac{1}{\pi} \int_0^\infty e^{-ty^2} y \sin xy \, dy. \qquad (3)$$

This step is valid since the integral (3) converges uniformly for $-\infty < x < \infty$, as one sees by the relation

$$\int_0^\infty e^{-ty^2} y \sin xy \, dy \ll \int_0^\infty e^{-ty^2} y \, dy = \frac{1}{2t}, \quad t > 0.$$

Integration by parts gives

$$u_x(x, t) = \frac{-x}{2\pi t} \int_0^\infty e^{-ty^2} \cos xy \, dy = \frac{-x}{2t} u(x, t).$$

Solving this differential equation, we obtain

$$u(x, t) = A(t) e^{-x^2/4t},$$

and we may determine $A(t)$ by setting $x = 0$:

$$u(0, t) = A(t) = \frac{1}{\pi} \int_0^\infty e^{-ty^2} \, dy = \frac{1}{\sqrt{4\pi t}}.$$

Here we have again used the value of the probability integral, and we have proved that $u(x, t) = k(x, t)$, as stated. The integral (2) is clearly equivalent to (1).

For the second proof, integrate e^{-s^2}, $s = \sigma + i\tau$, over the boundary of the rectangle

$$R = \{(\sigma, \tau) \mid -A \leq \sigma \leq A, \; 0 \leq \tau \leq B\}$$

in the complex s-plane. The result is zero by Cauchy's theorem, so that

$$\int_{-A}^{A} e^{-\sigma^2} \, d\sigma = \int_{0}^{B} e^{-(-A+i\tau)^2} \, d\tau - \int_{0}^{B} e^{-(A+i\tau)^2} \, d\tau + \int_{-A}^{A} e^{-(\sigma+iB)^2} \, d\sigma. \quad (4)$$

Clearly, each of the first two integrals on the right is dominated by

$$e^{-A^2} \int_{0}^{B} e^{\tau^2} \, d\tau = o(1), \quad A \to \infty.$$

Let $A \to \infty$ in (4) and use the value of the probability integral to obtain for all B

$$\sqrt{\pi} = \int_{-\infty}^{\infty} e^{-(\sigma+iB)^2} \, d\sigma = e^{B^2} \int_{-\infty}^{\infty} e^{-\sigma^2 - 2\sigma Bi} \, d\sigma.$$

Now make the change of variable $\sigma = -y\sqrt{t}$ and choose $B = x/\sqrt{4t}$:

$$\sqrt{\pi} = \sqrt{t} \, e^{x^2/(4t)} \int_{-\infty}^{\infty} e^{-ty^2 + xyi} \, dy.$$

This is equivalent to (2), and the proof is complete.

6 A FURTHER ADDITION FORMULA FOR $k(x,t)$

In §3, we derived an expression for $k(x, t_1 + t_2)$ in terms of $k(x, t_1)$ and $k(x, t_2)$. By use of Theorem 5, we can now obtain a related expression for $k(x, t_1 - t_2)$. Here is the result.

Theorem 6

1. $0 < t_1 < t_2, \quad -\infty < x, \quad y < \infty$

$$\Rightarrow \quad k(x - y, t_2 - t_1) = \int_{-\infty}^{\infty} k(r + ix, t_1) k(y - ir, t_2) \, dr. \quad (1)$$

The integral (1) is equal to

$$\frac{1}{4\pi\sqrt{t_1 t_2}} \exp\left(\frac{x^2}{4t_1} - \frac{y^2}{4t_2}\right)$$

$$\times \int_{-\infty}^{\infty} \exp\left[-\frac{(t_2 - t_1)r^2}{4t_1 t_2} + \frac{ir(t_1 y - t_2 x)}{2t_1 t_2}\right] dr.$$

But this integral can be evaluated by 5(2). Thus the integral (1) is equal to

$$\frac{1}{4\pi\sqrt{t_1 t_2}} \exp\left(\frac{x^2}{4t_1} - \frac{y^2}{4t_2}\right) k\left(\frac{t_1 y - t_2 x}{2t_1 t_2}, \frac{t_2 - t_1}{4t_1 t_2}\right).$$

From the definition of k, this is equal to $k(x - y, t_2 - t_1)$, as we wished to prove.

7 LAPLACE TRANSFORM OF $k(x, t)$

It will be useful to have the Laplace transform of $k(x, t)$ considered first as a function of x and then as a function of t.

Theorem 7.1

1. $s = \sigma + i\tau,\quad t > 0$

$$\Rightarrow \quad e^{ts^2} = \int_{-\infty}^{\infty} e^{-sy} k(y, t)\, dy, \quad -\infty < \sigma < \infty. \tag{1}$$

This result follows easily from Theorem 5. Representation 5(2) was derived for real x, but the integral clearly converges for all complex s when x is replaced by s. Hence the integral 5(2) serves to extend $k(x, t)$ analytically into the complex s-plane:

$$k(s, t) = \frac{1}{2\pi} \int_{-\infty}^{\infty} e^{-ty^2 + syi}\, dy, \quad t > 0,\ -\infty < \sigma < \infty. \tag{2}$$

Replace s by is and t by $1/(4t)$ to obtain

$$k\left(is, \frac{1}{4t}\right) = \frac{1}{2\pi} \int_{-\infty}^{\infty} e^{-sy} e^{-y^2/(4t)}\, dy = \sqrt{\frac{t}{\pi}} \int_{-\infty}^{\infty} e^{-sy} k(y, t)\, dy.$$

The conclusion now follows from the definition of $k(x, t)$.

Observe that we can also reclaim (2) from (1) by the usual inversion formula for the bilateral Laplace transform [Widder, 1946; 241]:

$$k(x, t) = \frac{1}{2\pi i} \int_{-\infty}^{i\infty} e^{sx} e^{s^2 t} \, ds = \frac{1}{2\pi} \int_{-\infty}^{\infty} e^{ixy - ty^2} \, dy.$$

Note also that Theorem 3 is now an obvious consequence of the product theorem for bilateral Laplace transforms [Widder, 1946; 258]. That is, the addition formula for k is the Laplace transform image of the addition formula for the exponential function:

$$e^{t_1 s^2} \cdot e^{t_2 s^2} = e^{(t_1 + t_2) s^2}.$$

Corollary 7.1

1. $t > 0$

$$\Rightarrow \quad -2s e^{ts^2} = \int_{-\infty}^{\infty} e^{-sy} h(y, t) \, dy, \quad -\infty < \sigma < \infty.$$

This follows from (1) by differentiation with respect to s, a step always valid for convergent Laplace integrals.

Since $k(x, t) \equiv 0$ for $t < 0$, its Laplace transform in the time variable is unilateral.

Theorem 7.2

1. $s = \sigma + i\tau, \quad -\infty < x < \infty$

$$\Rightarrow \quad \frac{e^{-|x|\sqrt{s}}}{\sqrt{4s}} = \int_0^{\infty} e^{-st} k(x, t) \, dt, \quad |\arg s| < \frac{\pi}{2}. \tag{3}$$

That branch of the square root which is real on the positive axis of reals is here intended. The integral (3) for $s = \sigma$ is

$$\frac{1}{\sqrt{4\pi}} \int_0^{\infty} e^{-\sigma t} \frac{e^{-x^2/(4t)}}{\sqrt{t}} \, dt = \frac{1}{\sqrt{\pi \sigma}} \int_0^{\infty} e^{-r^2 - (\sigma x^2/4 r^2)} \, dr, \quad r = \sqrt{\sigma t}.$$

The integral on the right is known [Widder, 1961c; 379] and yields equation (3) for $s = \sigma$. But the Laplace integral converges in the half-plane $\sigma > 0$ and defines an analytic function of s there. Hence the desired conclusion follows by analytic continuation.

8 LAPLACE TRANSFORM OF $h(x, t)$

We shall need also the Laplace integral of $h(x, t)$ considered as a function of t.

Theorem 8.1

1. $s = \sigma + i\tau$, $\quad 0 < x < \infty$

$$\Rightarrow \quad e^{-x\sqrt{s}} = \int_0^\infty e^{-st} h(x, t) \, dt, \quad |\arg s| < \frac{\pi}{2}. \tag{1}$$

We recall that $h(x, t) = -2k_x(x, t)$, so that we have only to differentiate equation 7(3) with respect to x to obtain (1) formally. Differentiation under the integral sign is valid since the resulting integral (1) converges uniformly in the interval $\delta \leqslant x < \infty$ for arbitrary $\delta > 0$. By use of Lemma 7, Chapter I, with $y = 1/t$, $p = \frac{3}{2}$, we see that

$$0 < h(x, t) < \frac{A}{x^2}, \quad x > 0, \ t > 0,$$

for a suitable constant A. Hence, for $\delta \leqslant x < \infty$

$$\int_0^\infty e^{-st} h(x, t) \, dt \ll \int_0^\infty e^{-\sigma t} \frac{A}{\delta^2} \, dt = \frac{A}{\delta^2 \sigma}, \quad \sigma > 0.$$

Since the dominant integral is independent of x, we have the desired uniform convergence of integral (1). Finally, we evaluate the Laplace transform of

$$l(x, t) = \operatorname{erfc}\left(\frac{x}{\sqrt{4t}}\right) = \frac{2}{\sqrt{\pi}} \int_{x/\sqrt{4t}}^\infty e^{-r^2} \, dr.$$

Theorem 8.2

1. $s = \sigma + i\tau$, $\quad 0 < x < \infty$

$$\Rightarrow \quad \frac{e^{-x\sqrt{s}}}{s} = \int_0^\infty e^{-st} l(x, t) \, dt, \quad |\arg s| < \frac{\pi}{2}. \tag{2}$$

Again it will be sufficient to prove the result for real $s = \sigma$. Evidently $l_x = -2k$, so that

$$l(x, t) = 2 \int_x^\infty k(y, t) \, dy, \quad x, t > 0.$$

The integral (2) becomes

$$2 \int_0^\infty e^{-\sigma t} \, dt \int_x^\infty k(y, t) \, dy = 2 \int_x^\infty dy \int_0^\infty e^{-\sigma t} k(y, t) \, dt$$

$$= \frac{1}{\sqrt{\sigma}} \int_x^\infty e^{-y\sqrt{\sigma}} \, dy = \frac{e^{-x\sqrt{\sigma}}}{\sigma}, \quad \sigma > 0. \tag{3}$$

The interchange in order of integration is valid since the integrand is positive. The inner integral on the right of (3) was evaluated by Theorem 7.2.

9 OPERATIONAL CALCULUS

Very useful as a guide to the understanding of the heat equation is a rudimentary knowledge of the operational calculus. In it, the procedure is to employ some operational symbol, such as D for differentiation, throughout some calculation as if it were a number and then finally to restore to it its original meaning. This procedure can be expected to produce meaningful results if the algebra employed in the calculations obeys the same basic laws (for example, commutative, distributive, and associative) as those enjoyed by the operator.

As an example, let us seek to give a meaning to the operator e^{aD}, where D stands for the operation of differentiation with respect to x and a is a constant. If D were a number we should have

$$e^{aD} = \sum_{n=0}^{\infty} \frac{a^n}{n!} D^n.$$

Applying the operator to a function $f(x)$ and interpreting D^n as an nth derivative, we have

$$e^{aD}f(x) = \sum_{n=0}^{\infty} \frac{a^n}{n!} f^{(n)}(x),$$

and this sum is $f(x + a)$, by Maclaurin's theorem, for suitable functions $f(x)$. Accordingly, we make the following definition, even if $f(x)$ has no known differentiability properties.

Definition 9.1

$$e^{aD}f(x) = f(x + a).$$

As a further example, let us try to attach a suitable meaning to $\exp(tD^2)$. If D were a number, the function

$$u(x, t) = e^{tD^2}f(x) \tag{1}$$

would satisfy the heat equation. For, then

$$u_{xx} = e^{tD^2}f''(x) = e^{tD^2}D^2f(x);$$

$$u_t = e^{tD^2}D^2f(x).$$

Moreover, $u(x, 0) = f(x)$, so that the function (1) should be a solution of Cauchy's problem. But we have yet to assign a meaning to the operator $\exp(tD^2)$.

Proceeding as above, by series expansion, we have

$$u(x, t) = e^{tD^2} f(x) = \sum_{n=0}^{\infty} \frac{t^n}{n!} D^{2n} f(x) = \sum_{n=0}^{\infty} \frac{t^n}{n!} f^{(2n)}(x). \qquad (2)$$

We see by direct differentiation that $u(x, t) \in H$ and that $u(x, 0) = f(x)$, as predicted. For example, if $f(x) = x^2$, the series has only two terms and $u(x, t) = v_2(x, t) = x^2 + 2t$, a heat polynomial. Or if $f(x) = \cos ax$, then $u(x, t) = e^{-a^2 t} \cos ax$.

An alternative approach to defining this operator is provided by the Laplace integral 7(1),

$$e^{tD^2} = \int_{-\infty}^{\infty} e^{-yD} k(y, t) \, dy,$$

since e^{-yD} has already been defined. Thus (1) becomes

$$u(x, t) = \int_{-\infty}^{\infty} e^{-yD} f(x) k(y, t) \, dy = \int_{-\infty}^{\infty} f(x - y) k(y, t) \, dy. \qquad (3)$$

That is, $u(x, t)$ is equal to the bilateral convolution which we met in §3 and which we called the Poisson transform. We already saw in §3 that this transform should also provide a solution of Cauchy's problem for suitable functions $f(x)$. For example, if $f(x) = e^{ax}$, then by 7(1)

$$u(x, t) = \int_{-\infty}^{\infty} k(y, t) e^{a(x-y)} \, dy = e^{ax + a^2 t}$$

$$= \sum_{n=0}^{\infty} \frac{t^n}{n!} D^n e^{ax}.$$

Thus (2) and (3) are equivalent for this special function $f(x)$. We adopt both definitions for the present.

Definition 9.2

$$e^{aD^2} f(x) = \sum_{n=0}^{\infty} \frac{a^n}{n!} f^{(2n)}(x).$$

Definition 9.3

$$e^{aD^2} f(x) = k(x, a) * f(x) = \int_{-\infty}^{\infty} k(x - y, a) f(y) \, dy.$$

It is evident that the two definitions do not have the same domain of applicability. The first is meaningful only if $f(x) \in C^\infty$ and if the series converges, the second can be applied to integrable functions for which the integral converges. We discuss later a class of functions $f(x)$ for which the two definitions are equivalent.

As a further example of the operational calculus, let $\mathcal{D}$ stand for differentiation with respect to t and seek to assign a meaning to $\cosh x\sqrt{\mathcal{D}}$. If $\mathcal{D}$ were a number, the function

$$u(x, t) = \cosh(x\sqrt{\mathcal{D}})f(t) \tag{4}$$

would satisfy the heat equation. Then

$$u_{xx} = \cosh(x\sqrt{\mathcal{D}})\mathcal{D}f(t) = \cosh(x\sqrt{\mathcal{D}})f'(t),$$

$$u_t = \cosh(x\sqrt{\mathcal{D}})f'(t).$$

Also

$$u(0, t) = f(t), \qquad u_x(0, t) = 0, \tag{5}$$

so that the function (4) should provide a solution to the Cauchy problem for the line $t = 0$ (not a characteristic), at least when one of the boundary functions is zero. The power series expansion for $\cosh x$ provides a suitable meaning for this operator.

Definition 9.4

$$\cosh(x\sqrt{\mathcal{D}})f(t) = \sum_{n=0}^{\infty} \frac{x^{2n}}{(2n)!} f^{(n)}(t).$$

The series formally provides a function of H satisfying (5).

As a final example, consider

$$u(x, t) = e^{-x\sqrt{\mathcal{D}}} f(t).$$

The series expansion would introduce fractional derivatives. A more useful definition for this operator is provided by the integral 7(3):

$$u(x, t) = e^{-x\sqrt{\mathcal{D}}} f(t) = \int_0^\infty e^{-y\mathcal{D}} f(t) h(x, y) \, dy$$

$$= \int_0^\infty f(t - y) h(x, y) \, dy.$$

If we assume that $f(t)$ vanishes for negative t, this integral reduces to a unilateral convolution,

$$u(x, t) = \int_0^t h(x, y) f(t - y)\, dy = h(x, y) * f(y). \tag{6}$$

This is the integral 6(2) of Chapter I, which we saw should define a temperature function. The operational calculus suggests that $u(0, t) = f(t)$ and the integral itself shows that $u(x, 0) = 0$. We study this integral in detail in Chapter IV.

10 THREE CLASSES OF FUNCTIONS

There are three classes of functions of the real variable which will be especially useful to us. Two of these were designated as "one" and "two" by Goursat [1923; 305], and we retain his notation for them while adding a third. The first is the familiar class of analytic functions for which we have already used the symbol A (for real or complex functions). Class II is less restrictive, admitting functions whose successive derivatives increase with the order more rapidly than those of analytic functions. Class III is more restrictive, demanding less rapid increase. Hence we have the inclusion relations

$$\text{III} \subset \text{I} \subset \text{II}$$

for the three classes. At times we may wish to introduce a parameter into the notation, in the interest of precision. In the following formal definitions, the constants M and r are independent of x and k, depending only on the function f.

Definition 10.1

$$f(x) \in \text{I}_r \quad \text{on} \quad a \leqslant x \leqslant b$$

$$\Leftrightarrow \quad |f^{(k)}(x)| < \frac{Mk!}{r^k}, \quad a \leqslant x \leqslant b, \ k = 0, 1, 2, \ldots. \tag{1}$$

Definition 10.2

$$f(x) \in \text{II}_r \quad \text{on} \quad a \leqslant x \leqslant b$$

$$\Leftrightarrow \quad |f^{(k)}(x)| < \frac{M(2k)!}{r^k}, \quad a \leqslant x \leqslant b, \ k = 0, 1, 2, \ldots. \tag{2}$$

10. THREE CLASSES OF FUNCTIONS

Definition 10.3

$$f(x) \in \text{III}_r \quad \text{on} \quad a \leq x \leq b$$

$$\Leftrightarrow \quad |f^{(2k)}(x)| < \frac{Mk!}{r^k}, \quad a \leq x \leq b, \; k = 0, 1, 2, \ldots,$$

$$|f^{(2k+1)}(x)| < \frac{Mk!}{r^k}, \quad a \leq x \leq b, \; k = 0, 1, 2, \ldots. \quad (3)$$

Implicit in these definitions is the assumption that $f(x) \in C^\infty$. Before giving examples let us recall the proof of the equivalence of classes A and I.

Theorem 10

$$f(x) \in \text{I} \Leftrightarrow f(x) \in \text{A} \quad \text{on} \quad a \leq x \leq b.$$

Assume first that $f(x) \in \text{I}$ on the interval J, $\{x | a \leq x \leq b\}$. To show that $f(x) \in A$, we show that it has a power series expansion valid in a neighborhood of an arbitrary point x_0 of J. By Maclaurin's theorem

$$f(x) = \sum_{k=0}^{n} f^{(k)}(x_0) \frac{(x - x_0)^k}{k!}$$

$$+ \frac{f^{(n+1)}(x_0 + \theta(x - x_0))(x - x_0)^{n+1}}{(n+1)!}, \quad \begin{array}{l} x \in J, \\ 0 < \theta < 1. \end{array}$$

The remainder is in absolute value $< M(|x - x_0|/r)^{n+1}$ by (1) and so tends to zero for $|x - x_0| < r$. Thus $f(x)$ has the required expansion.

Conversely, if $f(x) \in A$ on J, then its power series expansion for $x = x_0 \in J$,

$$f(x) = \sum_{k=0}^{\infty} f^{(k)}(x_0) \frac{(x - x_0)^k}{k!}, \quad (4)$$

may be used to define f in the complex z-plane, $z = x + iy$, merely by setting $x = z$ in (4). Thus $f(z)$ is analytic in some circle about each point x_0 of J. By use of the Heine–Borel theorem, in classic fashion, we may be assured that $f(z) \in A$ in some domain D including J in its interior. Let Γ be a regular curve in D, surrounding J, of length L, on which $|f(z)| < K$,

III. FURTHER DEVELOPMENTS

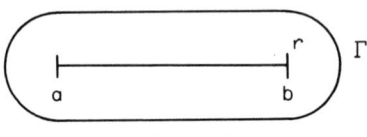

Figure 4

and whose minimum distance from J is r. (See Figure 4.) By Cauchy's integral,

$$f^{(k)}(x) = \frac{k!}{2\pi i} \int_\Gamma \frac{f(z)}{(z-x)^{k+1}} \, dz,$$

$$|f^{(k)}(x)| < \frac{k! \, KL}{2\pi r^{k+1}}, \qquad x \in J, \quad k = 0, 1, 2, \ldots.$$

This is equivalent to (1) and $f(x) \in I_r$ on J.

A function of class I is of course determined throughout its interval of definition by its values in a subinterval. This property does not obtain for functions of class II. We show in §11 that $e^{-1/x^2} \in \text{II}$ over any finite interval, if defined as 0 at $x = 0$. If its definition were altered to be $\equiv 0$ for $t < 0$, it would obviously remain in the class since all of its derivatives would still vanish at $x = 0$, and inequalities (4) would hold trivially for $x < 0$. Thus we have exhibited two different functions having equal values for $x > 0$.

11 EXAMPLES OF CLASS II

We show here, as corollaries of a general theorem, that the functions $k(x, t)$ and $h(x, t)$ as functions of t are in the class II for $-\infty < t < \infty$. Of course they are analytic in any interval not including $t = 0$.

Theorem 11.1

$$\begin{aligned}
&1. \ p > 0, \qquad\qquad\qquad m \geqslant 0, \\
&2. \ f(x) = x^{-m} e^{-p/x}, \qquad x > 0, \\
& = 0, \qquad\qquad\qquad x \leqslant 0,
\end{aligned}$$

$\Rightarrow \qquad f(x) \in \text{II}_r, \qquad -\infty < x < \infty, \quad r < 2p.$

For $x > 0$ we express $f(x)$ by Cauchy's integral formula, integrating over the circle (Figure 5),

$$\Gamma = \{z \mid z = x[1 + (1-\epsilon)e^{i\theta}]\}. \tag{1}$$

11. EXAMPLES OF CLASS II

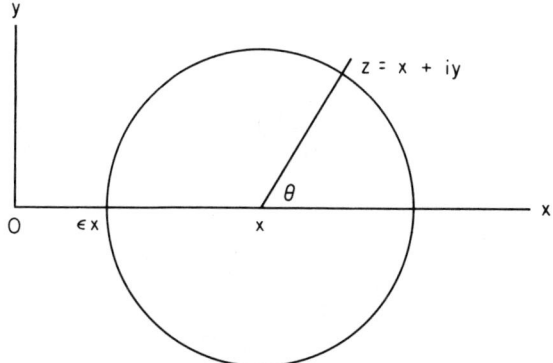

Figure 5

We obtain for $k = 0, 1, 2, \ldots$

$$f^{(k)}(x) = \frac{k!}{2\pi i} \int_\Gamma \frac{e^{-p/z}\, dz}{(z-x)^{k+1} z^m}.$$

Since the minimum value of $\operatorname{Re}(1/z)$ for $z \in \Gamma$ occurs at $\theta = 0$ we have

$$\operatorname{Re} \frac{1}{z} = \frac{1 + \rho \cos \theta}{x(1 + 2\rho \cos \theta + \rho^2)} \geq \frac{1}{x(\rho + 1)}, \qquad \rho = 1 - \epsilon,$$

so that

$$|f^{(k)}(x)| \leq \frac{k!\, e^{-p/x(\rho+1)}}{(x\rho)^k (1-\rho)^m x^m}.$$

Using 7(2) of Chapter I,

$$e^{-cy} y^n \leq \left(\frac{n}{ec}\right)^n, \qquad 0 \leq y < \infty, \tag{2}$$

with $n = k + m$, $c = p/(\rho + 1)$, and $y = 1/x$, we have

$$|f^{(k)}(x)| \leq \frac{k!}{\rho^k (1-\rho)^m} \left[\frac{(k+m)(\rho+1)}{ep}\right]^{k+m}, \qquad 0 \leq x < \infty. \tag{3}$$

This inequality also holds trivially when $x < 0$. To conclude the proof we have only to show that the right-hand side of (3) is less than $M(2k)!/r^k$ for suitable constants M and r. This will follow if

$$\lim_{k \to \infty} \frac{r^k}{M(2k)!} \frac{k!}{\rho^k (1-\rho)^m} \left[\frac{(k+m)(\rho+1)}{ep}\right]^{k+m} = 0. \tag{4}$$

But the function of k on the left is the kth term of a series whose test ratio is

$$\lim_{k \to \infty} \frac{rk}{(2k)(2k-1)} \frac{1}{\rho} \frac{\rho+1}{e\rho} (k+m)\left(1 + \frac{1}{k+m-1}\right)^{k+m-1}$$
$$= \frac{r(\rho+1)}{4p\rho},$$

and the series converges if

$$r < \frac{4p\rho}{\rho+1}.$$

Since $e^{-p/z}$ is singular only at $z = 0$, we may choose ρ as near to 1 as we like, so that (2) is established when $r < 2p$, and the theorem is proved.

Corollary 11.1

1. $a \neq 0$

$$\Rightarrow \quad k(a, t), h(a, t) \in \mathrm{II}_r, \quad r < \frac{a^2}{2}, \quad -\infty < t < \infty.$$

Apply the theorem with $p = a^2/4$ and $m = \frac{1}{2}$ for $k(x, t)$, $m = \frac{3}{2}$ for $h(x, t)$. Another useful function in the class II is e^{-1/x^2}.

Theorem 11.2

1. $f(x) = e^{-1/x^2}, \quad x \neq 0;$

$\quad f(0) = 0;$

$\Rightarrow \quad f(x) \in \mathrm{II}_r, \quad \text{every } r > 0, \quad -\infty < x < \infty.$

We again use Cauchy's integral over the circle (1) with $\epsilon = \rho = \frac{1}{2}$. Then for z on Γ

$$\mathrm{Re}\,\frac{1}{z^2} = \frac{1 + \cos\theta + \frac{1}{4}\cos 2\theta}{x^2(\frac{5}{4} + \cos\theta)^2} \geq \frac{\frac{1}{4}}{x^2(\frac{9}{4})^2} = \frac{\delta}{x^2}.$$

Hence

$$f^{(k)}(x) = \frac{k!}{2\pi i} \int_\Gamma \frac{e^{-1/z^2}\,dz}{(z-x)^{k+1}}, \quad x > 0,$$

$$|f^{(k)}(x)| \leq k!\left(\frac{2}{x}\right)^k e^{-\delta/x^2} \leq k!\, 2^k \left(\frac{k}{2e\delta}\right)^{k/2}. \tag{5}$$

Here we have used (2) with $y = x^{-2}$, $c = \delta$, $n = k/2$. To show that the right-hand side of (5) is less than $M(2k)!/r^k$ for every $r > 0$, it will suffice to show that

$$\lim_{k \to \infty} \frac{r^k k!}{(2k)!} \left(\frac{2k}{e\delta} \right)^{k/2} = 0,$$

or that the test ratio

$$\lim_{k \to \infty} \frac{rk}{2k(2k-1)} \left(\frac{2k}{\delta e} \right)^{1/2} \left(1 + \frac{1}{k-1} \right)^{(k-1)/2} < 1.$$

But this limit is clearly zero for all r. We have proved that $f(x) \in \text{II}$ for $0 \leq x < \infty$, but by symmetry the result holds for $-\infty < x < \infty$.

12 RELATION AMONG THE CLASSES

In §10, we indicated the inclusion relations among the class I–III. Here we give three examples to substantiate those claims.

Example A The function $f(x)$ of Theorem 11.2 serves to distinguish between classes I and II. By that theorem, it belongs to II. That it cannot belong to I follows since e^{-1/z^2} has a singularity at the origin of the complex z-plane. Another proof follows directly from Definition 10.1. Suppose $f(x)$ did belong to I and that inequalities 10(1) held. Since all derivatives of $f(x)$ vanish at the origin we have by Taylor's formula for $x < r$

$$f(x) = f^{(k)}(X) \frac{x^k}{k!}, \qquad 0 < X < x,$$

$$|f(x)| < M \left(\frac{x}{r} \right)^k = o(1), \qquad k \to \infty.$$

That is, $f(x) \equiv 0$ on $(0, r)$, a contradiction.

Example B $f(x) = (1/x) \in \text{I}_\delta$ on $\delta \leq x < \infty$ for any $\delta > 0$; for

$$|f^{(k)}(x)| = \frac{k!}{x^{k+1}} \leq \frac{k!}{\delta^{k+1}}, \qquad \delta \leq x < \infty.$$

But $f(x) \notin \text{III}$. For if it did, there would exist constants M and r such that

$$f^{(2k)}(x) = \frac{(2k)!}{x^{2k+1}} \leq \frac{Mk!}{r^k}, \qquad x \geq \delta, \quad k = 0, 1, 2, \ldots .$$

But this is impossible for any x since
$$\lim_{k \to \infty} \frac{(2k)!}{k!} \frac{r^k}{x^{2k+1}} = \infty.$$

Example C
$$f(x) = e^x \in \text{III} \quad \text{on} \quad |x| \leq R, \quad \text{every } R.$$

This follows since
$$f^{(2k)}(x) = f^{(2k+1)}(x) = e^x \leq e^R, \quad |x| \leq R.$$

Hence,
$$f^{(2k)}(x), f^{(2k+1)}(x) \leq \frac{Mk!}{r^k}, \quad k = 0, 1, 2, \ldots,$$

for $M = e^R$ and $r = 1$, for example.

The inclusion relations may be visualized by the Venn diagram of Figure 6.

It is clear that any polynomial belongs to class III on any finite interval; for, all derivatives vanish after a certain order and inequalities 10(3) are satisfied trivially for a suitable constant M.

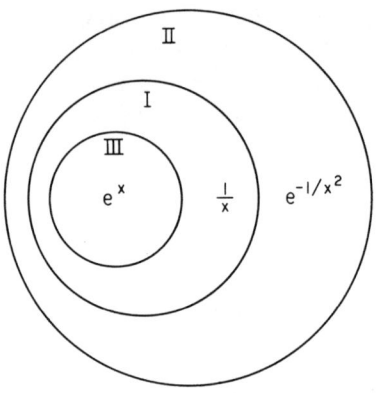

Figure 6

13 SERIES EXPANSIONS OF FUNCTIONS IN CLASS I

In §9, we conjectured that
$$u(x, t) = \cosh x\sqrt{\mathcal{D}}\, \varphi(t), \quad \mathcal{D} = \frac{d}{dt},$$

would provide a solution of Cauchy's problem for the line $x = 0$,
$$u(0, t) = \varphi(t), \quad u_x(0, t) = \psi(t),$$

13. SERIES EXPANSIONS OF FUNCTIONS IN CLASS I

in the special case when $\psi(t) \equiv 0$. Here we treat the general case when φ and ψ are arbitrary functions of class I.

Theorem 13

$\quad$ 1. $\varphi(t), \psi(t) \in I, \quad a \leqslant t \leqslant b,$

$\quad$ 2. $u(x, t) = \sum_{k=0}^{\infty} \varphi^{(k)}(t) \dfrac{x^{2k}}{(2k)!} + \sum_{k=0}^{\infty} \psi^{(k)}(t) \dfrac{x^{(2k+1)}}{(2k+1)!}$ $\quad$ (1)

$\Rightarrow$ $\quad$ A. $\quad u(x, t) \in H, \quad\quad a \leqslant t \leqslant b, \quad -\infty < x < \infty,$

$\quad\quad$ B. $\quad u(0, t) = \varphi(t), \quad\quad a \leqslant t \leqslant b,$

$\quad\quad$ C. $\quad u_x(0, t) = \psi(t), \quad\quad a \leqslant t \leqslant b.$

Formal differentiation of series (1) gives

$$u_x(x, t) = \sum_{k=0}^{\infty} \varphi^{(k+1)}(t) \dfrac{x^{2k+1}}{(2k+1)!} + \sum_{k=0}^{\infty} \psi^{(k)}(t) \dfrac{x^{2k}}{(2k)!} \quad (2)$$

$$u_{xx}(x, t) = u_t(x, t) = \sum_{k=0}^{\infty} \varphi^{(k+1)}(t) \dfrac{x^{2k}}{(2k)!} + \sum_{k=0}^{\infty} \psi^{(k+1)}(t) \dfrac{x^{2k+1}}{(2k+1)!}. \quad (3)$$

The validity of these operations follows from the uniform convergence of series (2) and (3). By 1, the functions φ and ψ satisfy inequalities 10(1), so that for $|x| \leqslant R$ and $a \leqslant t \leqslant b$

$$\sum_{k=0}^{\infty} \varphi^{(k+1)}(t) \dfrac{x^{2k+1}}{(2k+1)!} \ll M \sum_{k=0}^{\infty} \dfrac{(k+1)!}{(2k+1)!} \dfrac{R^{2k+1}}{r^{k+1}} < \infty. \quad (4)$$

Hence the series on the left of (4) converges uniformly for $|x| \leqslant R$ and $a \leqslant t \leqslant b$ by the Weierstrass M-test. Similarly, the other three series, (2)(3), converge uniformly in the same rectangle. Since R may be taken arbitrarily large, conclusion A is established. Conclusions B and C follow by direct substitution.

It is noteworthy that the analyticity of φ and ψ [however small r may be in 10(1)] implies that $u(z, t)$, $z = x + iy$, is an entire function of z for each fixed t. Observe further that series (1) serves to extend any temperature function, defined only in a neighborhood of a segment of the t-axis and satisfying 1 there, into a complete horizontal strip containing that segment.

Example A

$$\varphi(t) = e^{b^2 t}, \qquad \psi(t) \equiv 0.$$

Here φ and $\psi \in I$ on $-\infty < t < \infty$, and

$$\cosh x\sqrt{\mathcal{D}}\;\varphi(t) = \sum_{k=0}^{\infty} e^{b^2 t} b^{2k} \frac{x^{2k}}{(2k)!} = e^{b^2 t} \cosh bx.$$

This function $\in H$ over the whole plane and is entire in z when x is replaced by the complex variable z.

14 SERIES EXPANSIONS OF FUNCTIONS IN CLASS II

The series 13(1) still serves to solve Cauchy's problem when the prescribed functions belong to class II, but the solution is generally defined in a smaller region.

Theorem 14

1. $\varphi(t), \psi(t) \in II_r, \qquad a \leqslant t \leqslant b,$
2. $u(x, t) = \sum_{k=0}^{\infty} \varphi^{(k)}(t) \dfrac{x^{2k}}{(2k)!} + \sum_{k=0}^{\infty} \psi^{(k)}(t) \dfrac{x^{2k+1}}{(2k+1)!}$

$\Rightarrow$
A. $u(x, t) \in H, \qquad a \leqslant t \leqslant b, \quad -r < x < r,$
B. $u(0, t) = \varphi(t), \qquad a \leqslant t \leqslant b,$
C. $u_x(0, t) = \psi(t), \qquad a \leqslant t \leqslant b.$

The relation 13(4) is now replaced by

$$\sum_{k=0}^{\infty} \varphi^{(k+1)}(t) \frac{x^{2k+1}}{(2k+1)!} \ll M \sum_{k=0}^{\infty} \frac{(2k+2)!}{(2k+1)!} \frac{R^{2k+1}}{r^{2k+2}}.$$

The dominant series converges for $R < r$, so that the desired conclusion follows as in the previous proof. The series for ψ is treated similarly.

Example A

$$\varphi(t) = \frac{e^{-1/t}}{\sqrt{t}}, \qquad t > 0,$$

$$= 0, \qquad t \leqslant 0,$$

$$\psi(t) = 0, \qquad -\infty < t < \infty.$$

Thus, $\varphi(t) = \sqrt{4\pi}\, k(2, t)$. By Theorem 5,

$$\varphi(t) = \frac{2}{\sqrt{\pi}} \int_0^\infty e^{-ty^2} \cos 2y \, dy, \qquad t > 0.$$

Hence,

$$u(x, t) = \frac{2}{\sqrt{\pi}} \int_0^\infty e^{-ty^2} \sum_{k=0}^\infty (-1)^k \frac{(xy)^{2k}}{(2k)!} \cos 2y \, dy$$

if term-by-term integration is valid. It is so since

$$\int_0^\infty e^{-ty^2} \sum_{k=0}^\infty \frac{|xy|^{2k}}{(2k)!} \, dy = \int_0^\infty e^{-ty^2} \cosh xy \, dy < \infty, \qquad t > 0.$$

Here we have used a familiar criterion. See Titchmarsh [1939; 45]. Consequently,

$$u(x, t) = \frac{2}{\sqrt{\pi}} \int_0^\infty e^{-ty^2} \cos xy \cos 2y \, dy$$

$$= \frac{1}{\sqrt{\pi}} \int_0^\infty e^{-ty^2} [\cos(x+2)y + \cos(x-2)y] \, dy$$

$$= \sqrt{\pi} \, [k(x+2, t) + k(x-2, t)] = \frac{e^{-(x^2+4)/(4t)}}{\sqrt{t}} \cosh\!\left(\frac{x}{t}\right), \qquad t > 0,$$

$$= 0, \qquad t \leq 0. \tag{1}$$

Note that $u(z, t)$ is an entire function of the complex variable z for *every* t, $-\infty < t < \infty$. However, $u(x, t) \in H$ in the vertical strip $|x| < 2$ but in no larger vertical strip. For, $u(x, t)$ clearly has singularities at $(\pm 2, 0)$. For example, $u(2, 0+) = +\infty$. By Corollary 11.1, $\varphi(t) \in \mathrm{II}_r$, $r < 2$, so that $u(x, t) \in H$ in the strip predicted by the theorem. Thus, Theorem 14 cannot be improved in this respect, as another example in §16 will show. It is interesting to observe that the function (1) is the Appell transform of the more familiar function $\sqrt{4\pi}\, e^t \cosh x$.

15 SERIES EXPANSIONS OF FUNCTIONS IN CLASS III

Functions in class III serve well as initial functions for Cauchy's problem on a characteristic, at least if the solution is to be obtained by a power series.

Theorem 15.1

1. $f(x) \in \text{III}_r, \quad a \leqslant x \leqslant b,$

2. $u(x, t) = e^{tD^2}f(x) = \sum_{k=0}^{\infty} f^{(2k)}(x) \frac{t^k}{k!}$ (1)

$\Rightarrow$ A. $u(x, t) \in H, \quad -r < t < r, \quad a \leqslant x \leqslant b,$

B. $u(x, 0) = f(x).$

By use of inequalities 10(3), we can show that series (1) converges uniformly in $a \leqslant x \leqslant b$, $-r + \delta \leqslant t \leqslant r - \delta$ for any positive $\delta < r$. Indeed we may show the same for the differentiated series; for

$$\sum_{k=0}^{\infty} f^{(2k+2)}(x) \frac{t^k}{k!} \ll M \sum_{k=0}^{\infty} \frac{(k+1)!}{k!} \left(\frac{r-\delta}{r}\right)^k < \infty. \quad (2)$$

Since the dominant series is independent of x and t we may use the Weierstrass M-test to show that series (2) converges uniformly for $a \leqslant x \leqslant b$, $-r + \delta \leqslant t \leqslant r - \delta$ and hence has the sum $u_{xx}(x, t)$ or $u_t(x, t)$ there. Thus A is established and B follows trivially.

It is noteworthy that $u(x, t)$ is defined for negative t, a circumstance not to be forseen from physical considerations. It is so here on account of the stringent conditions imposed on $f(x)$ for admission into class III. Indeed for functions in this class we can show that Definitions 9.2 and 9.3 of e^{aD^2} are equivalent for $a > 0$.

Theorem 15.2

1. $f(x) \in \text{III}_r, \quad -\infty < x < \infty,$

$\Rightarrow \quad \int_{-\infty}^{\infty} k(x - y, t) f(y) \, dy = \int_{-\infty}^{\infty} k(y, t) f(x - y) \, dy$

$$= \sum_{k=0}^{\infty} f^{(2n)}(x) \frac{t^n}{n!},$$

$-\infty < x < \infty, \quad 0 < t < r.$

Since $\text{III} \subset \text{I}$, $f(x) \in A$ and Maclaurin's theorem yields

$$f(x - y) = \sum_{n=0}^{\infty} f^{(n)}(x) \frac{(-y)^n}{n!} \quad (3)$$

15. SERIES EXPANSIONS OF FUNCTIONS IN CLASS III

valid for any fixed x and small y. However, inequalities 10(3) show that series (3) converges for all y, so that (3) is valid for all x, y by analytic continuation. Hence,

$$\int_{-\infty}^{\infty} k(y, t) f(x - y)\, dy = \int_{-\infty}^{\infty} k(y, t) \sum_{n=0}^{\infty} f^{(n)}(x) \frac{(-y)^n}{n!} \, dy$$

$$= \sum_{n=0}^{\infty} \frac{(-1)^n}{n!} f^{(n)}(x) \int_{-\infty}^{\infty} k(y, t) y^n \, dy \quad (4)$$

if term-by-term integration is valid. It is so if

$$\sum_{n=0}^{\infty} \frac{|f^{(n)}(x)|}{n!} \int_{-\infty}^{\infty} k(y, t) |y|^n \, dy < \infty \quad (5)$$

by the criterion used in §14 [Titchmarsh, 1939; 45]. By inequalities 10(3), series (5) is dominated by

$$M \sum_{n=0}^{\infty} \frac{n!}{(2n)!} \frac{1}{r^n} \int_{-\infty}^{\infty} k(y, t) y^{2n} \, dy$$

$$+ M \sum_{n=0}^{\infty} \frac{n!}{(2n+1)!} \frac{1}{r^n} \int_{-\infty}^{\infty} k(y, t) |y|^{2n+1} \, dy. \quad (6)$$

Setting $y^2 = z$, we have

$$\int_{-\infty}^{\infty} k(y, t) y^{2n} \, dy = \frac{1}{\sqrt{4\pi t}} \int_0^{\infty} e^{-z/(4t)} z^{n-(1/2)} \, dz$$

$$= \frac{(4t)^{n+(1/2)}}{\sqrt{4\pi t}} \Gamma(n + \tfrac{1}{2}) < \frac{(4t)^n n!}{\sqrt{\pi}}$$

$$\int_{-\infty}^{\infty} k(y, t) |y|^{2n+1} \, dy = \frac{1}{\sqrt{4\pi t}} \int_0^{\infty} e^{-z/(4t)} z^n \, dz = \frac{(4t)^{n+1} n!}{\sqrt{4\pi t}}. \quad (7)$$

Hence series (6) will converge if

$$\sum_{n=0}^{\infty} \frac{n!\, n!}{(2n)!} \left(\frac{4t}{r}\right)^n + \sum_{n=0}^{\infty} \frac{n!\, n!}{(2n+1)!} \frac{(4t)^{n+1}}{r^n} < \infty.$$

The test ratios for these two series are

$$\lim_{n \to \infty} \frac{n^2}{(2n)(2n-1)} \frac{4t}{r} = \lim_{n \to \infty} \frac{n^2}{(2n+1)(2n)} \frac{4t}{r} = \frac{t}{r}.$$

Hence (4) is established if $0 < t < r$. Since the integral (4) is zero when n is odd we have by (7) that

$$\int_{-\infty}^{\infty} k(y, t) f(x - y) \, dy = \sum_{n=0}^{\infty} \frac{f^{(2n)}(x)}{(2n)!} \int_{-\infty}^{\infty} k(y, t) y^{2n} \, dy$$

$$= \frac{1}{\sqrt{\pi}} \sum_{n=0}^{\infty} \frac{f^{(2n)}(x)}{(2n)!} (4t)^n \Gamma(n + \tfrac{1}{2}).$$

But

$$\Gamma(n + \tfrac{1}{2}) = (n - \tfrac{1}{2})(n - \tfrac{3}{2}) \cdots \tfrac{1}{2}\sqrt{\pi} = \frac{(2n)!\sqrt{\pi}}{n! \, 4^n},$$

so that the theorem is proved. Note that the Poisson integral definition of $e^{tD^2}f(x)$ never provides a temperature function for negative t (because of the factor $1/\sqrt{4\pi t}$), whereas the series definition may do so. As noted before, the integral is much more generally applicable in other respects.

Example A.

$$f(x) = x^{2n}, \qquad n \text{ an integer} > 0.$$

Then by Theorem 15.1

$$f^{(2k)}(x) = \frac{(2n)! \, x^{2n-2k}}{(2n - 2k)!}, \qquad k = 0, 1, \ldots, n;$$

$$u(x, t) = (2n)! \sum_{k=0}^{n} \frac{x^{2n-2k}}{(2n - 2k)!} \frac{t^k}{k!} = v_{2n}(x, t).$$

This is the heat polynomial defined in Chapter I, §5. Since $x^{2n} \in \text{III}$, Theorem 15.2 gives

$$\int_{-\infty}^{\infty} k(x - y, t) y^{2n} \, dy = v_{2n}(x, t).$$

In particular,

$$\int_{-\infty}^{\infty} k(y, t) y^{2n} \, dy = v_{2n}(0, t) = \frac{(2n)! \, t^n}{n!}.$$

In a similar way $k(x, t) * x^n = v_n(x, t)$ when n is odd.

15. SERIES EXPANSIONS OF FUNCTIONS IN CLASS III

Example B.
$$f(x) = e^{-x^2}.$$

This function $\in \text{III}_r$ on $-\infty < x < \infty$ when $r < \tfrac{1}{4}$; for,

$$f^{(n)}(x) = (-1)^n H_n(x) e^{-x^2},$$

where $H_n(x)$ is the Hermite polynomial of degree n. But it is known [Erdélyi, 1953; 208] that

$$|H_n(x)| < A e^{x^2/2} 2^{n/2} \sqrt{n!}, \qquad -\infty < x < \infty,$$

for some constant A. Hence,

$$|f^{(2n)}(x)| < A e^{-x^2/2} 2^n \sqrt{(2n)!} < \frac{Mn!}{r^n}, \qquad -\infty < x < \infty,$$

for suitable constants M and r. Such constants can certainly be found if

$$\lim_{n \to \infty} \frac{2^n r^n \sqrt{(2n)!}}{n!} = 0. \tag{8}$$

But the test ratio of the series for which this function of n is the nth term is

$$\lim_{n \to \infty} \frac{2r}{n} \sqrt{(2n)(2n-1)} = 4r.$$

Hence (8) is established when $4r < 1$. In a similar way one can obtain bounds on the odd order derivatives of $f(x)$, as required by Definition 8.3. Hence, $f(x) \in \text{III}_r$, $r < \tfrac{1}{4}$, as stated. Consequently the solution of the Cauchy problem is

$$u(x, t) = \sum_{n=0}^{\infty} D^{2n} e^{-x^2} \frac{t^n}{n!}, \qquad |t| < \tfrac{1}{4}. \tag{9}$$

To find $u(x, t)$ explicitly, we appeal to Theorem 15.2,

$$u(x, t) = \int_{-\infty}^{\infty} k(x - y, t) e^{-y^2} \, dy = \sqrt{\pi}\, k(x, t) * k(x, \tfrac{1}{4}), \qquad t > 0,$$

$$= \sqrt{\pi}\, k(x, t + \tfrac{1}{4}) = \frac{e^{-x^2/(4t+1)}}{\sqrt{4t+1}}.$$

Here we have used Theorem 3. In the complex s-plane, $k(x, s + \frac{1}{4})$ has a singularity at $s = -\frac{1}{4}$ so that the power series (9) must have radius of convergence $\frac{1}{4}$. In this way, we see that Theorem 15.1 is the best possible in a certain sense.

16 A TEMPERATURE FUNCTION WHICH IS NOT ENTIRE IN THE SPACE VARIABLE

All solutions $u(x, t)$ which we have thus far encountered have been restrictions to the real axis x of functions $u(z, t)$, $z = x + iy$, which were entire functions of z for fixed t. By use of Theorem 14, we give an example of a function which does not have this property. In that theorem, choose $\psi(t) \equiv 0$ and

$$\varphi(t) = \int_0^\infty e^{-y} \cos(ty^2)\, dy.$$

Clearly

$$|\varphi^{(n)}(t)| \leq \int_0^\infty e^{-y} y^{2n}\, dy = (2n)!, \qquad -\infty < t < \infty,$$

so that $\varphi(t) \in \mathrm{II}_1$ on $-\infty < t < \infty$. Then the function

$$u(x, t) = \sum_{n=0}^\infty \varphi^{(n)}(t) \frac{x^{2n}}{(2n)!}$$

belongs to H for $-\infty < t < \infty$, $-1 < x < 1$. Since

$$\varphi^{(2n)}(0) = (-1)^n \int_0^\infty e^{-y} y^{4n}\, dy = (-1)^n (4n)! \tag{1}$$

$$\varphi^{(2n+1)}(0) = 0, \qquad n = 0, 1, 2, \ldots,$$

we have

$$u(x, 0) = \sum_{n=0}^\infty (-1)^n x^{4n} = \frac{1}{1 + x^4}, \qquad -1 < x < 1.$$

The analytic extension of this function clearly has singularities on the unit circle, so that $u(z, 0)$ is not entire.

This example shows also that Theorem 14 is best possible. That is, the conclusion A depends *precisely* on hypothesis 1. We cannot hope to prove

16. A TEMPERATURE FUNCTION NOT ENTIRE IN THE SPACE VARIABLE

that $u(x, t) \in H$ in any wider strip than $|x| < r$. For the present example, $\varphi(t) \notin \mathrm{II}_r$ for any $r > 1$, since the inequality

$$|\varphi^{(2n)}(t)| < \frac{M(4n)!}{r^n}$$

would lead to a contradiction through equation (1):

$$1 < \frac{M}{r^n}, \qquad n = 0, 1, 2, \ldots.$$

This is false for large n, however large M may be.

It may be shown easily that $u(x, t)$ can be given an integral expression as follows:

$$u(x, t) = \operatorname{Re} \int_0^\infty e^{-y - ity^2} \cosh xy \sqrt{i}\ dy.$$

Chapter IV

INTEGRAL TRANSFORMS

1 POISSON TRANSFORMS

In Chapter III, §3, we introduced the integral transform

$$u(x, t) = \int_{-\infty}^{\infty} k(x - y, t)\varphi(y) \, dy, \qquad (1)$$

calling it the *Poisson transform*. Here we study it in detail under general hypotheses. Without further statement we assume that $\varphi(y) \in L$ (Lebesgue integrable) in every finite interval. More generally, we sometimes wish to consider the Stieltjes integral

$$u(x, t) = \int_{-\infty}^{\infty} k(x - y, t) \, d\alpha(y), \qquad (2)$$

where $\alpha(y) \in V$ (bounded variation) in every finite interval. When we wish to distinguish between (1) and (2) we call the first the *Poisson–Lebesgue* transform, the second the *Poisson–Stieltjes* transform. We prove at once in the simplest case the fact conjectured earlier, that the function (1) solves the Cauchy problem for the characteristic $t = 0$ with boundary function $\varphi(x)$.

Theorem 1

1. $\varphi(x) \in B \cdot C$ (bounded, continuous) $-\infty < x < \infty$,
2. $u(x, t) = \int_{-\infty}^{\infty} k(x - y, t)\varphi(y)\, dy,$ $t > 0$,

$\Rightarrow$ $u(x, 0+) = \varphi(x),$ $-\infty < x < \infty.$ (3)

As in Chapter III, §3,

$$u(x, t) = \frac{1}{\sqrt{\pi}} \int_{-\infty}^{\infty} e^{-r^2}\varphi(x - r\sqrt{4t})\, dr$$

$$\ll \frac{M}{\sqrt{\pi}} \int_{-\infty}^{\infty} e^{-r^2}\, dr < \infty, \qquad M = \underset{-\infty<x<\infty}{\text{u.b.}} |\varphi(x)|.$$

By Lebesgue's limit theorem [Titchmarsh, 1939; 345], we may let $t \to 0+$ under the integral sign to obtain

$$u(x, 0+) = \frac{1}{\sqrt{\pi}} \int_{-\infty}^{\infty} e^{-r^2}\varphi(x)\, dr = \varphi(x), \qquad -\infty < x < \infty.$$

That $u(x, t) \in H$ for $t > 0$ will follow later, under more general hypotheses.

From this result, the physical meaning of $u(x, t)$ becomes clear. It is the temperature of an infinite bar supplied with a linear coordinate at a point x of the bar t seconds after it was $\varphi(x)$.

For later purposes, it will be useful to know that the limit (3) is uniform on finite intervals.

Corollary 1

1. $R > 0$

$\Rightarrow$ $\lim_{t \to 0+} u(x, t) = \varphi(x),$ uniformly on $|x| \leq R.$

Given $\epsilon > 0$, we show that for some δ

$$|u(x, t) - \varphi(x)| < \epsilon, \qquad 0 < t < \delta, \quad |x| \leq R.$$

Choose S so large that

$$\frac{1}{\sqrt{\pi}} \int_{|y|>S} e^{-y^2}|\varphi(x - y\sqrt{4t}) - \varphi(x)|\, dy \leq \frac{2M}{\sqrt{\pi}} \int_{|y|>S} e^{-y^2}\, dy < \frac{\epsilon}{2}.$$

Now choose δ so small that when $|x| \leq R$, $t < \delta$, $|y| \leq S$

$$|\varphi(x - y\sqrt{4t}) - \varphi(x)| < \frac{\epsilon}{2}.$$

This is possible by the uniform continuity of $\varphi(x)$ in the closed interval $|x| \leq 2R$, say. Hence

$$|u(x, t) - \varphi(x)| \leq \frac{1}{\sqrt{\pi}} \int_{-\infty}^{\infty} e^{-y^2} |\varphi(x - y\sqrt{4t}) - \varphi(x)| \, dy$$

$$< \frac{\epsilon}{2} + \frac{1}{\sqrt{\pi}} \int_{-S}^{S} e^{-y^2} \frac{\epsilon}{2} \, dy < \epsilon.$$

This concludes the proof.

2 CONVERGENCE

We show here that the region of convergence of the integral (2) is a horizontal strip, $0 < t < c$, of the x, t-plane. The same is true of the integral (1) to which (2) reduces when $\alpha(y)$ is an integral of $\varphi(y)$. We need the following preliminary result.

Lemma 2

1. $\alpha(x) \in V$, $\qquad a \leq x \leq R$, $\qquad$ every $R > a$,
2. $\alpha(+\infty)$ $\qquad\qquad$ exists,
3. $\varphi(x) \geq 0, \in C, \in \downarrow,$ $\qquad a \leq x < \infty,$

$\Rightarrow \qquad \displaystyle\int_a^\infty \varphi(x) \, d\alpha(x) \qquad$ converges.

Given $\epsilon > 0$, we determine T so large and $> a$, that

$$|\alpha(R) - \alpha(S)| < \epsilon, \qquad R, S > T.$$

This is possible by 2. Then

$$\int_R^S \varphi(x) \, d\alpha(x) = \int_R^S \varphi(x) \, d[\alpha(x) - \alpha(R)]$$

$$= \varphi(S)[\alpha(S) - \alpha(R)] - \int_R^S [\alpha(x) - \alpha(R)] \, d\varphi(x).$$

Hence,

$$\left| \int_R^S \varphi(x) \, d\alpha(x) \right| < \epsilon\varphi(S) + \epsilon[\varphi(R) - \varphi(S)] \leq \epsilon\varphi(a).$$

2. CONVERGENCE

By Cauchy's criterion for the existence of a limit, the desired conclusion follows.

Theorem 2

1. $\int_{-\infty}^{\infty} k(x_0 - y, t_0) \, d\alpha(y) = A,$
2. $0 < t < t_0,$ $\qquad -\infty < x < \infty$

$\Rightarrow$

A. $\int_{-\infty}^{\infty} k(x - y, t) \, d\alpha(y) \quad$ converges, $\qquad$ (1)

B. $\lim_{t \to t_0-} \int_{-\infty}^{\infty} k(x_0 - y, t) \, d\alpha(y) = A.$

Set

$$\beta(y) = \int_0^y e^{-(x_0 - r)^2/(4t_0)} \, d\alpha(r),$$

$$\varphi(y) = \exp\left[\frac{(x_0 - y)^2}{4t_0} - \frac{(x - y)^2}{4t} \right],$$

so that

$$\frac{1}{\sqrt{4\pi t}} \int_0^\infty \varphi(y) \, d\beta(y) = \int_0^\infty k(x - y, t) \, d\alpha(y). \qquad (2)$$

Moreover, $\varphi(y) \in C$, ≥ 0, and $\in \downarrow$ for sufficiently large y. Hence the integral (2) converges by Lemma 2. The integral

$$\int_{-\infty}^0 k(x - y, t) \, d\alpha(y)$$

also converges, as one sees by replacing y by $-y$ and applying the previous result.

To obtain B, we note that the integral (1) can be related to a Laplace integral as follows:

$$\int_{-\infty}^{\infty} k(x_0 - y, t) \, d\alpha(y)$$

$$= \frac{1}{\sqrt{4\pi t}} \int_0^\infty e^{-y/(4t)} \, d\big[\alpha(x_0 + \sqrt{y}) - \alpha(x_0 - \sqrt{y}) \big]. \qquad (3)$$

This integral is known to converge at $t = t_0$. Hence by a familiar Abelian theorem for such integrals [Widder, 1946; 56], conclusion B follows.

That the region of convergence for (1) is a horizontal strip in the x, t-plane now follows in a familiar way. Compare Widder [1946; 37].

3 POISSON TRANSFORM IN H

We show now that $u(x, t)$ defined by the integral 1(2) belongs to H in its strip of convergence.

Theorem 3

1. $u(x, t) = \int_{-\infty}^{\infty} k(x - y, t) \, d\alpha(y)$ converges at (x_0, t_0) (1)

$\Rightarrow \quad u(x, t) \in H, \quad 0 < t < t_0.$

Since $k(x, t) \in H$ for $t > 0$, we have only to justify differentiation under the integral sign, both with respect to x and with respect to t. This is done most simply by an appeal to the theory of the Laplace transform [Widder, 1946; 57]. By 2(3) it is clear that the integral (1) can be differentiated under the sign with respect to t as often as desired. As a function of x, the integral (1) may be related to a bilateral Laplace integral as follows:

$$u(x, t) = k(x, t) \int_{-\infty}^{\infty} e^{xy/(2t) - y^2/(4t)} \, d\alpha(y), \tag{2}$$

and again unlimited differentiation under the sign is valid [Widder, 1946; 57].

4 ANALYTICITY

In the space variable x, $u(x, t)$ as defined by 1(2) is the restriction to reals of a function which is an entire function of the complex variable.

Theorem 4.1

1. $u(x, t) = \int_{-\infty}^{\infty} k(x - y, t) \, d\alpha(y)$ converges at (x_0, t_0), $t_0 > 0$,
2. $s = \sigma + i\tau,$ $\quad 0 < t_1 < t_0$

$\Rightarrow \quad u(s, t_1) \in A \quad \text{(analytic)}, \quad |s| < \infty.$

By equation 3(2),

$$u(s, t_1) = k(s, t_1) \int_{-\infty}^{\infty} e^{sy/(2t_1) - y^2/(4t_1)} \, d\alpha(y),$$

the integral converging for $-\infty < \sigma < \infty$ by Theorem 2. Since it is a Laplace integral, it therefore represents an entire function of s. The factor $k(s, t_1)$ is also entire so that the theorem is proved.

Considered as a function of t, the transform $u(x, t)$ can also be extended analytically into the complex plane, though it will not generally be entire. We show that the extension is analytic in the disk for which the interval $(0, t_0)$ is a diameter.

Theorem 4.2

1. $u(x, t) = \int_{-\infty}^{\infty} k(x - y, t) \, d\alpha(y)$ converges at (x_0, t_0), $t_0 > 0$,
2. $s = \sigma + i\tau$, $-\infty < x_1 < \infty$

$\Rightarrow$ $u(x_1, s) \in A$, $\operatorname{Re} \dfrac{1}{s} > \dfrac{1}{t_0}$.

By equation 2(3),

$$u(x_1, s) = \frac{1}{\sqrt{4\pi s}} \int_0^{\infty} e^{-y/(4s)} \, d\big[\alpha(x_1 + \sqrt{y}) - \alpha(x_1 - \sqrt{y})\big], \quad (1)$$

the integral converging for $0 < \sigma < t_0$ by Theorem 2. Since a unilateral Laplace integral represents an analytic function in its half-plane of convergence, the integral (1) represents an analytic function for $\operatorname{Re}(1/s) > 1/t_0$,

$$\operatorname{Re} \frac{1}{s} = \frac{\sigma}{\sigma^2 + \tau^2} > \frac{1}{t_0},$$

$$\left(\sigma - \frac{t_0}{2}\right)^2 + \tau^2 < \frac{t_0^2}{4}. \quad (2)$$

Since $s^{-1/2} \in A$ in the same circle, the theorem is proved. Inequality (2) defines a disk with center at $(t_0/2, 0)$, radius $t_0/2$, so that the region of analyticity is as described above.

5 INVERSION OF THE POISSON–LEBESGUE TRANSFORM

By Theorem 1, we inverted the transform 1(1) when $\varphi(y)$ was bounded and continuous. We wish now to relax those hypotheses as far as possible.

We assume only that the transform converges at some point and that $\varphi(y)$ satisfies a mild local condition at the point where inversion is to be accomplished. We recall that $\varphi(y)$ is always assumed to be Lebesgue integrable in every finite interval.

Lemma 5

1. $u(x, t) = \int_a^\infty k(x - y, t)\varphi(y)\, dy$ converges at (x_0, t_0), $t_0 > 0$,
2. $\int_a^x [\varphi(y) - \varphi(a)]\, dy = o(x - a)$, $x \to a +$,

$\Rightarrow \qquad u(a, 0+) = \dfrac{\varphi(a)}{2}.$

For arbitrary $\delta > 0$, set

$$J_1(t) = \frac{1}{\sqrt{4\pi t}} \int_a^{a+\delta} \beta(y)\, d\alpha(y), \qquad J_2(t) = \frac{1}{\sqrt{4\pi t}} \int_{a+\delta}^\infty \beta(y)\, d\alpha(y),$$

where

$$\beta(y) = \exp\left[\frac{(y - x_0)^2}{4t_0} - \frac{(y - a)^2}{4t}\right], \tag{1}$$

$$\alpha(y) = \int_{a+\delta}^y e^{-(r - x_0)^2/(4t_0)} \varphi(r)\, dr,$$

so that

$$u(a, t) = J_1(t) + J_2(t).$$

To evaluate $J_1(0+)$, we employ Laplace's asymptotic method [Widder, 1946; 278], Theorem 2b, with $k = 1/t$ and $h(y) = -(y - a)^2/4$. Thus $h'(a) = 0$, $h''(a) = -\frac{1}{2}$, and $J_1(0+) = \varphi(a)/2$. We show next that $J_2(0+) = 0$. Since for $t < t_0$ it is the second of the exponents (1) that is dominant, we have $\beta(\infty) = 0$, and for t small, $\beta(y) \in \downarrow$ on $(a + \delta, \infty)$. There exists an upper bound M for $|\alpha(y)|$ on that interval since $\alpha(y) \in C$ there and $|\alpha(\infty)| < \infty$ by hypothesis 1. Applying integration by parts to $J_2(t)$, we have

$$J_2(t) = -\frac{1}{\sqrt{4\pi t}} \int_{a+\delta}^\infty \alpha(y)\, d\beta(y),$$

$$|J_2(t)| \leq \frac{M\beta(a + \delta)}{\sqrt{4\pi t}} = \frac{M}{\sqrt{4\pi t}} \exp\left[\frac{(a + \delta - x_0)^2}{4t_0} - \frac{\delta^2}{4t}\right].$$

5. INVERSION OF THE POISSON–LEBESGUE TRANSFORM

Since $e^{-\delta^2/(4t)}/\sqrt{t} \to 0$ as $t \to 0+$, we have $J_2(0+) = 0$, as stated. This concludes the proof.

Theorem 5

1. $u(x, t) = \int_{-\infty}^{\infty} k(x - y, t)\varphi(y)\, dy$ converges at (x_0, t_0),
2. $\int_a^x [\varphi(y) - \varphi(a)]\, dy = o(|x - a|)$, $x \to a$,

$\Rightarrow$ $u(a, 0+) = \varphi(a)$.

From hypothesis 1 we have

$$u(a, t) = \int_a^{\infty} k(a - y, t)\varphi(y)\, dy + \int_{-a}^{\infty} k(-a - y, t)\varphi(-y)\, dy. \quad (2)$$

To the first of these integrals we apply Lemma 5 directly to obtain the limit $\varphi(a)/2$ as $t \to 0+$. The lemma may also be applied to the second integral after appropriate changes in notation, using $x \to a-$ in the present hypothesis 2. The limit is again $\varphi(a)/2$ so that the theorem is proved.

Corollary 5a Hypothesis 1 with hypothesis 2 replaced by the following: $\varphi(a+)$ and $\varphi(a-)$ exist

$\Rightarrow$ $u(a, 0+) = \dfrac{\varphi(a+) + \varphi(a-)}{2}.$

Under these conditions,

$$\int_a^x [\varphi(y) - \varphi(a+)]\, dy = o(x - a), \quad x \to a+,$$

$$\int_x^a [\varphi(y) - \varphi(a-)]\, dy = o(x - a), \quad x \to a-,$$

and the two integrals may be treated separately by Lemma 5 to obtain the stated result.

Corollary 5b Hypothesis 1 with hypothesis 2 omitted

$\Rightarrow$ $u(a, 0+) = \varphi(a)$ almost all a.

It is known that for any Lebesgue integrable function $\varphi(x)$, hypothesis 2 holds for almost all a. See [Titchmarsh, 1939; 362], for example.

6 INVERSION OF THE POISSON–STIELTJES TRANSFORM

Here we again need some preliminary results.

Lemma 6a

1. $\int_{-\infty}^{\infty} k(x - y, t)\, d\alpha(y)$ converges at (x_0, t_0), $t_0 > 0$,

$\Rightarrow$ $\alpha(x) = o\left(\dfrac{1}{k(x - x_0, t_0)}\right)$, $|x| \to \infty$.

It will suffice to treat the case $x \to +\infty$. Given $\epsilon > 0$, by virtue of hypothesis 1 and Cauchy's criterion we can determine b so large and $> x_0$ that for any x_1 and x greater than b,

$$\left|\int_{x_1}^{x} k(x_0 - y, t_0)\, d\alpha(y)\right| < \epsilon. \tag{1}$$

Since $1/k(y - x_0, t_0) > 0$ and $\in \uparrow$ for $y > x_0$, we may apply the second law of the mean to obtain for some $\xi \geqslant x_1$

$$\alpha(x) - \alpha(x_1) = \int_{x_1}^{x} \frac{k(x_0 - y, t_0)}{k(x_0 - y, t_0)}\, d\alpha(y)$$

$$= \frac{1}{k(x - x_0, t_0)} \int_{\xi}^{x} k(x_0 - y, t_0)\, d\alpha(y).$$

By (1),

$$|\alpha(x)| k(x - x_0, t_0) \leqslant |\alpha(x_1)| k(x - x_0, t_0) + \epsilon.$$

Consequently,

$$\varlimsup_{x \to +\infty} |\alpha(x)| k(x - x_0, t_0) \leqslant \epsilon.$$

Since ϵ was arbitrary, this nonnegative limit superior must be zero, and the lemma is proved.

6. INVERSION OF THE POISSON–STIELTJES TRANSFORM

Lemma 6b

1. $\int_{-\infty}^{\infty} k(x - y, t) \, d\alpha(y)$ converges at (x_0, t_0), $t_0 > 0$,
2. $0 < t < t_0$, $0 < R < \infty$

$\Rightarrow \int_{-\infty}^{\infty} \frac{\partial k}{\partial x}(x - y, t)\alpha(y) \, dy$ converges uniformly on $|x| \leqslant R$.

Computing $\partial k/\partial x$, we have only to show that

$$\int_{-\infty}^{\infty} (x - y)k(x - y, t)\alpha(y) \, dy \qquad (2)$$

is dominated by a convergent integral independent of x on $|x| \leqslant R$. By Lemma 6a there exists a constant M such that

$$|\alpha(y)| \leqslant \frac{M}{k(y - x_0, t_0)}, \qquad -\infty < y < \infty.$$

Hence, for $|x| \leqslant R$ the integral (2) is dominated by

$$M \int_{-\infty}^{\infty} (|y| + R) \frac{k(x - y, t)}{k(x_0 - y, t_0)} \, dy$$

$$\ll M \sqrt{\frac{t_0}{t}} \int_{-\infty}^{\infty} (|y| + R) \exp\left[-\frac{(|y| - R)^2}{4t} + \frac{(y - x_0)^2}{4t_0}\right] dy.$$

Since the integrand of the latter integral is

$$O\left[|y| \exp\left(\frac{1}{4t_0} - \frac{1}{4t}\right) y^2\right], \qquad |y| \to \infty,$$

the integral converges, as desired.

Theorem 6

1. $u(x, t) = \int_{-\infty}^{\infty} k(x - y, t) \, d\alpha(y)$ converges at (x_0, t_0), $t_0 > 0$,
2. $-\infty < a < b < \infty$

$\Rightarrow \lim_{t \to 0+} \int_a^b u(x, t) \, dx = \frac{\alpha(b +) + \alpha(b -)}{2} - \frac{\alpha(a +) + \alpha(a -)}{2}.$

We recall our convention that α is assumed of bounded variation on finite intervals. Integrating by parts and observing that the integrated part vanishes by Lemma 6a, we have

$$u(x, t) = \int_{-\infty}^{\infty} \frac{\partial k}{\partial x}(x - y, t)\alpha(y)\, dy, \qquad 0 < t < t_0.$$

This integral converges uniformly on $a \leqslant x \leqslant b$ by Lemma 6b, so that

$$\int_a^b u(x, t)\, dx = \int_{-\infty}^{\infty} k(b - y, t)\alpha(y)\, dy - \int_{-\infty}^{\infty} k(a - y, t)\alpha(y)\, dy.$$

Since $\alpha(y) \in V$, it has right- and left-hand limits at every point, and we may apply Corollary 5a to obtain the desired limit.

7 THE h-TRANSFORM

In 9(6) of Chapter III, we introduced the unilateral convolution $h(x, t) * \varphi(t)$,

$$u(x, t) = \int_0^t h(x, t - y)\varphi(y)\, dy, \tag{1}$$

$$h(x, t) = \frac{x}{t} k(x, t) = \frac{xe^{-x^2/(4t)}}{(4\pi t^3)^{1/2}}, \qquad t > 0,$$

$$= 0, \qquad t \leqslant 0.$$

We define equation (1) as the *h-transform*. We shall sometimes need to consider the more general *h-Stieltjes transform*

$$u(x, t) = \int_0^t h(x, t - y)\, d\alpha(y)$$

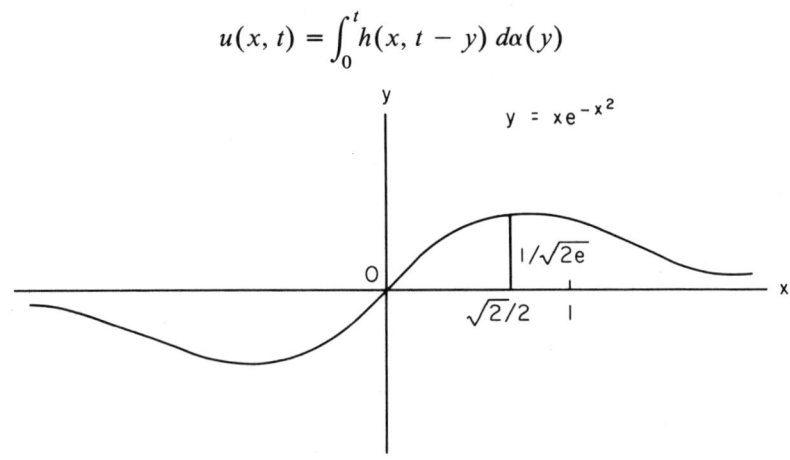

Figure 7

which of course includes (1) when $\alpha(y)$ is the integral of $\varphi(y)$. To study it in detail we need some preliminary facts. A graph of $(\sqrt{\pi}/4)h(x, \frac{1}{4})$ is given in Figure 7.

Theorem 7.1

A. $h(x, t) > 0$, $\qquad x, t > 0$;
B. $h(x, 0+) = h(x, +\infty) = 0$, $\qquad x > 0$;
C. $h(x_0, t) \in \uparrow$, $\qquad x_0 > 0, \ 0 < t < x_0^2/6$;
    $h(x_0, t) \in \downarrow$, $\qquad x_0 > 0, \ x_0^2/6 < t < \infty$;
D. $\int_0^\infty h(x, t) \, dt = 1$, $\qquad x > 0$;
E. $\lim_{x \to 0+} \int_0^c h(x, t) \, dt = 1$, $\qquad c > 0$.

These facts are easily proved, the last from the equation

$$\int_0^c h(x, t) \, dt = \frac{2}{\sqrt{\pi}} \int_{x/\sqrt{4c}}^\infty e^{-y^2} \, dy.$$

Theorem 7.2

A. $\lim_{x \to 0+} \int_c^\infty |h_t(x, t)| \, dt = 0$, $\qquad c > 0$;
B. $\int_0^\infty t|h_t(x, t)| \, dt < 2$, $\qquad x > 0$.

By C of Theorem 7.1, we have for $0 < x < \sqrt{6c}$

$$\int_c^\infty |h_t(x, t)| \, dt = -\int_c^\infty h_t(x, t) \, dt = h(x, c) \to 0, \qquad x \to 0+.$$

Similarly,

$$\int_0^\infty t|h_t(x, t)| \, dt = \int_0^{x^2/6} th_t(x, t) \, dt - \int_{x^2/6}^\infty th_t(x, t) \, dt$$

$$= 2 \frac{x^2}{6} h\left(x, \frac{x^2}{6}\right) - \int_0^{x^2/6} h(x, t) \, dt + \int_{x^2/6}^\infty h(x, t) \, dt$$

$$< \frac{\sqrt{6}}{\sqrt{\pi}} e^{-3/2} + 1 < 2.$$

Here we have integrated by parts and used D of the previous theorem.

We now prove the basic property of the h-transform (1) in the simplest case, $\varphi \in C$.

Theorem 7.3

    1. $\varphi(x) \in C,$ $0 \leqslant x < \infty,$

    2. $u(x, t) = \int_0^t h(x, t - y)\varphi(y)\, dy$

$\Rightarrow$ A. $u(0+, t) = \varphi(t),$ $0 < t < \infty,$

        B. $u(x, 0+) = 0,$ $0 < x < \infty.$

By change of variable,

$$u(x, t) = \frac{2}{\sqrt{\pi}} \int_{x/\sqrt{4t}}^{\infty} e^{-r^2} \varphi\left(t - \frac{x^2}{4r^2}\right) dr \ll \frac{2M_t}{\sqrt{\pi}} \int_0^{\infty} e^{-r^2}\, dr = M_t,$$

$$M_t = \max_{0 \leqslant y \leqslant t} |\varphi(y)|.$$

By Lebesgue's limit theorem [Titchmarsh, 1939; 345],

$$u(0+, t) = \frac{2}{\sqrt{\pi}} \int_0^{\infty} e^{-r^2} \varphi(t)\, dr = \varphi(t),$$

and A is proved. From the above,

$$|u(x, t)| \leqslant \frac{2M_t}{\sqrt{\pi}} \int_{x/\sqrt{4t}}^{\infty} e^{-r^2}\, dr, \quad x > 0.$$

The right-hand side clearly $\to 0$ as $t \to 0+$, and B is established.

This theorem shows that $u(x, t)$ as defined by (1) may be interpreted as the temperature of a semi-infinite bar. It is extended along a positive x-axis; its temperature is everywhere 0 at $t = 0$; its temperature at the finite end, $x = 0$, is prescribed at time t as $\varphi(t)$. Then $u(x, t)$ is the temperature of the bar at point x at time t.

If we add to the hypotheses of Theorem 7.3 the condition $\varphi(0+) = 0$, we can show that the limits in A and B are uniform.

Corollary 7.3

              3. $\varphi(0+) = 0$

$\Rightarrow$ A. $\lim\limits_{x \to 0+} u(x, t) = \varphi(t)$ uniformly on $0 \leqslant t \leqslant c;$

        B. $\lim\limits_{t \to 0+} u(x, t) = 0$ uniformly on $0 \leqslant x < \infty.$

We prove B first. Given ϵ, determine δ by 3 such that

$$|\varphi(t)| < \epsilon, \qquad 0 < t < \delta.$$

Hence for $0 < t < \delta$,

$$|u(x, t)| \leq \int_0^t h(x, t - y)|\varphi(y)| \, dy < \epsilon \int_0^\infty h(x, t - y) \, dy = \epsilon. \qquad (2)$$

Since δ is independent of x, B is proved.

Given ϵ, we seek to determine η independent of t on $0 \leq t \leq c$ such that when $0 < x < \eta$

$$|u(x, t) - \varphi(t)| < \epsilon, \qquad 0 \leq t \leq c. \qquad (3)$$

We draw the conclusion in two ways corresponding to the subintervals $0 \leq t \leq \delta$ and $\delta \leq t \leq c$, where δ is now determined so that when $0 < y < \delta$, we have $|\varphi(y)| < \epsilon/2$ and

$$|\varphi(t - y) - \varphi(t)| < \epsilon/2, \qquad 0 \leq t \leq c, \; 0 \leq t - y \leq c. \qquad (4)$$

This is possible by (3) and the uniform continuity of $\varphi(t)$ on $0 \leq t \leq c$. On $0 \leq t \leq \delta$ we have $|\varphi(t)| < \epsilon/2$ and we may apply (2) (replacing ϵ by $\epsilon/2$). Hence (3) holds on that interval for *all* x. On $\delta \leq t \leq c$ we use E of Theorem 7.1 as follows:

$$u(x, t) - \varphi(t) = \int_0^t h(x, y)\varphi(t - y) \, dy - \varphi(t) \int_0^\infty h(x, y) \, dy;$$

$$|u(x, t) - \varphi(t)| \leq \int_0^t h(x, y)|\varphi(t - y) - \varphi(t)| \, dy + |\varphi(t)| \int_t^\infty h(x, y) \, dy$$

$$\leq \frac{\epsilon}{2} \int_0^\infty h(x, y) \, dy + M \int_\delta^\infty h(x, y) \, dy,$$

$$M = \max_{0 \leq t \leq c} |\varphi(t)|.$$

Again using Theorem 7.1, we now determine η so that $0 < x < \eta$ implies

$$M \int_\delta^\infty h(x, y) \, dy < \epsilon/2.$$

Consequently, (3) also holds on (δ, c) and A is proved.

Without the addition of hypothesis 3 the uniformity would fail, as the example $\varphi(t) \equiv 1$ shows. Then

$$u(x, t) = \frac{2}{\sqrt{\pi}} \int_{x/\sqrt{4t}}^{\infty} e^{-y^2} dy = \mathrm{erfc}\left(\frac{x}{\sqrt{4t}}\right).$$

Here $u(x, 0+) = 0$ for $x > 0$, $u(0+, t) = 1$ for $t > 0$, and $u(0+, 0) = 0$, so that both conclusions of the corollary are violated.

8 h-TRANSFORM IN H

Like the Poisson transform, the h-transform defines a temperature function. We treat the general Stieltjes case.

Theorem 8

1. $\alpha(y) \in V$, $\quad 0 \leq y \leq R$, every $R > 0$,
2. $u(x, t) = \int_0^t h(x, t - y) \, d\alpha(y)$

$\Rightarrow \quad u(x, t) \in H, \quad 0 < x < \infty, \quad -\infty < t < \infty.$

It is no restriction to assume $\alpha(0) = 0$. Integration by parts gives

$$u(x, t) = \int_0^t h_t(x, t - y) \alpha(y) \, dy.$$

Since for any $\delta > 0$, $h(x, t) \in C^\infty$ in the rectangle $\delta \leq x \leq R$ and $-R \leq t \leq R$, we may apply classical theorems to obtain in that rectangle

$$u_{xx}(x, t) = \int_0^t h_{txx}(x, t - y) \alpha(y) \, dy. \tag{1}$$

If $\alpha(t)$ were continuous,

$$u_t(x, t) = \int_0^t h_{tt}(x, t - y) \alpha(y) \, dy + \alpha(t) h_t(x, 0),$$

the last term vanishing. We prove that this is still true for $\alpha(t) \in V$. For $\delta > 0$ and $t > 0$, we have

$$\frac{u(x, t + \delta) - u(x, t)}{\delta}$$

$$= \frac{1}{\delta} \int_0^t [h_t(x, t + \delta - y) - h_t(x, t - y)] \alpha(y) \, dy$$

$$+ \frac{1}{\delta} \int_t^{t+\delta} h_t(x, t + \delta - y) \alpha(y) \, dy.$$

The last term tends to 0 with δ, as predicted, since it is dominated by

$$|h_t(x, \theta\delta)| \cdot \underset{0 \leq t \leq R}{\text{u.b.}} |\alpha(t)|, \qquad 0 < \theta < 1,$$

and $h_t(x, 0) = 0$. The first term tends to

$$u_t(x, t) = \int_0^t h_{tt}(x, t - y)\alpha(y)\, dy \tag{2}$$

by Lebesgue's limit theorem, for example. This equation holds trivially for $t \leq 0$ where h vanishes identically. Since $h_{txx} = h_{tt}$ when $(x, t) \neq (0, 0)$ the integrals (1) and (2) are equal and the theorem is proved.

Corollary 8

$$u(x, t) \in C^\infty, \qquad 0 < x < \infty, \quad -\infty < t < \infty.$$

This follows, as above, by successive differentiations under the integral sign.

9 ANALYTICITY

We show now that $u(x, t)$, as defined by an h-transform, belongs to class I in the space variable and to class II in the time variable.

Theorem 9.1

1. $u(x, t) = \int_0^t h(x, t - y)\, d\alpha(y), \qquad x > 0,$
2. $s = \sigma + i\tau, \qquad t_0 > 0,$

$\Rightarrow \qquad u(s, t_0) \in A, \qquad |\arg s| < \pi/4.$

The change of variable $y = t_0 - (4r)^{-1}$ gives

$$u(s, t_0) = \frac{4s}{\sqrt{\pi}} \int_{1/(4t_0)}^\infty e^{-s^2 r} r^{3/2}\, d\left[-\alpha\left(t_0 - \frac{1}{4r}\right)\right] dr. \tag{1}$$

This is a Laplace integral in which one variable has been replaced by s^2. Since the h-transform is well defined for $x > 0$, the Laplace integral converges in a right half-plane and the integral (1) represents an analytic function for

$$\text{Re } s^2 = \sigma^2 - \tau^2 > 0$$

and hence surely in the angular region $|\arg s| < \pi/4$.

Corollary 9.1

1. $0 < a < b,$ $0 < t_0,$

$\Rightarrow \quad u(x, t_0) \in \mathrm{I}, \quad a \leq x \leq b.$

This follows from Theorem 10, Chapter III.

Theorem 9.2

1. $u(x, t) = \int_0^t h(x, t - y)\, d\alpha(y), \quad x > 0,$
2. $x_0 > 0,$ $c > 0$

$\Rightarrow \quad u(x_0, t) \in \mathrm{II}, \quad -c \leq t \leq c.$

By Theorem 8 and its corollary,

$$\frac{\partial^n u}{\partial t^n} = \frac{\partial^{2n} u}{\partial x^{2n}}, \quad 0 < x < \infty, \quad -\infty < t < \infty. \tag{2}$$

By Corollary 9.1,

$$\left| \frac{\partial^{2n} u}{\partial x^{2n}} \right| \leq \frac{M(2n)!}{r^{2n}}, \quad n = 0, 1, \ldots,$$

for some constants M and r, which may depend on t_0. If we can show them independent of t_0 on $(-\infty, \infty)$, then the desired conclusion will follow

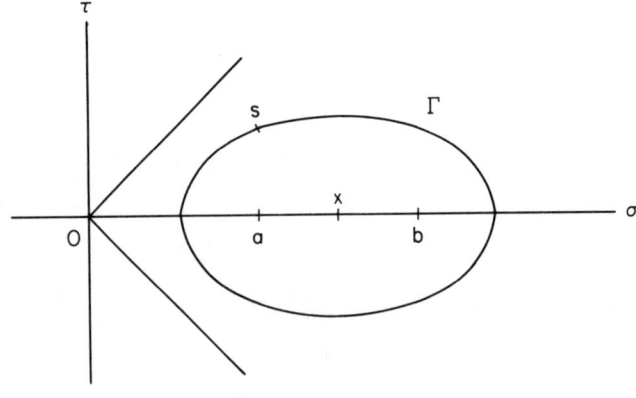

Figure 8

from equation (2). To do so we first show that there exist constants M and r independent of x on $a \leq x \leq b$ and independent of t on $-\infty < t < \infty$, such that

$$\left| \frac{\partial^n h(x, t)}{\partial x^n} \right| < \frac{Mn!}{r^n}. \tag{3}$$

Construct a closed curve Γ of length L, in the complex s-plane, bounding a region including the interval $a \leq \sigma \leq b$ and lying in the angular region $|\arg s| < \pi/4$, Figure 8. Since Γ touches neither the rays $\arg s = \pm \pi/4$ nor the interval $a \leq \sigma \leq b$, there exist positive constants, δ, Δ, r such that for $s \in \Gamma$ and $a \leq x \leq b$,

$$\operatorname{Re} s^2 = \sigma^2 - \tau^2 \geq \delta, \qquad |s - x| \geq r, \qquad |s| \leq \Delta,$$

and so

$$|h(s, t)| = \frac{|s|}{\sqrt{4\pi}} \frac{e^{-(\sigma^2 - \tau^2)/(4t)}}{t^{3/2}} \leq \frac{\Delta e^{-\delta^2/(4t)}}{\sqrt{4\pi}\, t^{3/2}} < B, \qquad 0 < t < \infty.$$

Here B is a constant which could be obtained explicitly from Lemma 7, Chapter I. By Cauchy's integral formula

$$h^{(n)}(x, t) = \frac{n!}{2\pi i} \int_\Gamma \frac{h(s, t)}{(s - x)^{n+1}} \, ds,$$

$$|h^{(n)}(x, t)| \leq \frac{n!\, BL}{2\pi r^{n+1}}, \qquad 0 < t < \infty.$$

This inequality is trivial for $t \leq 0$, so that (3) is established.

Since all derivatives of $h(x, t)$ vanish at $t = 0$, when $x > 0$ we have, as in §8,

$$\frac{\partial^n u}{\partial t^n} = \int_0^t \frac{\partial^n h}{\partial t^n} (x, t - y) \, d\alpha(y) = \int_0^t \frac{\partial^{2n} h}{\partial x^{2n}} (x, t - y) \, d\alpha(y).$$

If V is the total variation of $\alpha(y)$ on $0 \leq y \leq c$, then from (3)

$$\left| \frac{\partial^n u}{\partial t^n} (x, t) \right| \leq \frac{M(2n)!\, V}{r^{2n}}, \qquad a \leq x \leq b, \quad -c \leq t \leq c,$$

and the theorem is proved.

The function $u(x, t)$ of these two theorems is not, in general, an analytic function of t. A special case is $h(x, t)$ itself which is not analytic at $t = 0$, $x \neq 0$.

10 INVERSION OF THE h-LEBESGUE TRANSFORM

By Theorem 7.3 we inverted this transform in a special case. We treat now the general case. As usual, $\varphi(y)$ is assumed to be Lebesgue integrable in every finite interval.

Theorem 10.1

1. $u(x, t) = \int_0^t h(x, t - y)\varphi(y)\, dy, \qquad t > 0,$
2. $\int_0^x [\varphi(t_0 - y) - \varphi(t_0)]\, dy = o(x), \qquad x \to 0+, \quad t_0 > 0,$

$\Rightarrow \qquad u(0+, t_0) = \varphi(t_0).$

The result is immediate when $\varphi(y)$ is constant by Theorem 7.1. Hence we need only show that

$$\lim_{x \to 0+} \int_0^{t_0} h(x, y)[\varphi(t_0 - y) - \varphi(t_0)]\, dy = 0.$$

Set

$$\alpha(x) = \int_0^x [\varphi(t_0 - y) - \varphi(t_0)]\, dy,$$

so that

$$\alpha(x) = o(x), \qquad x \to 0+ \quad \text{by 2,}$$

and

$$u(x, t_0) - \varphi(t_0) = \int_0^{t_0} h(x, y)\, d\alpha(y)$$

$$= \alpha(t_0) h(x, t_0) - \int_0^{t_0} h_y(x, y)\alpha(y)\, dy.$$

The first term $\to 0$ as $x \to 0$. We break the integral into two others, I_1 and I_2, corresponding, respectively, to the intervals $(0, \delta)$ and (δ, t_0), $\delta > 0$. By Theorem 7.2,

$$|I_1| \leq \int_0^\delta y|h_y(x, y)|\frac{|\alpha(y)|}{y}\, dy \leq 2 \max_{0 < y \leq \delta} \frac{|\alpha(y)|}{y}.$$

10. INVERSION OF THE h-LEBESGUE TRANSFORM

Also

$$|I_2| \leq \max_{0 \leq y \leq t_0} |\alpha(y)| \int_\delta^{t_0} |h_y(x,y)| \, dy = o(1), \qquad x \to 0+,$$

by Theorem 7.2. Hence

$$\varlimsup_{x \to 0+} |u(x,t_0) - \varphi(t)| \leq 2 \max_{0 \leq y \leq \delta} \frac{|\alpha(y)|}{y}.$$

The left-hand side being independent of δ must be 0 since the right-hand side $\to 0$ as $\delta \to 0$ by hypothesis 2. The desired conclusion now follows immediately.

Corollary 10.1a If hypothesis 2 is omitted, $u(0+, t) = \varphi(t)$ almost everywhere.

Since for any Lebesgue integrable function hypothesis 2 is automatically satisfied almost everywhere, the conclusion follows.

Corollary 10.1b If hypothesis 2 is replaced by $\varphi(t_0 -) = \varphi(t_0)$, then $u(0+, t_0) = \varphi(t_0)$.

The present hypothesis implies hypothesis 2 of the theorem.

Theorem 10.2

$$\begin{aligned}&1. \quad u(x,t) = \int_0^t h(x, t-y) \, d\alpha(y), \qquad t > 0, \\ &2. \quad \alpha(0) = 0\end{aligned}$$

$$\Rightarrow \qquad \alpha(t-) = \lim_{x \to 0+} \int_0^t u(x,y) \, dy, \qquad t > 0.$$

Integration by parts and further obvious calculations give

$$u(x,t) = \int_0^t h_t(x, t-y) \alpha(y) \, dy,$$

$$\int_0^t u(x,y) \, dy = \int_0^t dr \int_0^r h_r(x, r-y) \alpha(y) \, dy$$

$$= \int_0^t \alpha(y) \, dy \int_y^t h_r(x, r-y) \, dr$$

$$= \int_0^t h(x, t-y) \alpha(y) \, dy.$$

Since $\alpha(y)$ is of bounded variation, $\alpha(t-)$ exists and we may apply Corollary 10.1b to obtain the desired conclusion.

11 THE k-TRANSFORM

We use this designation for the equation

$$u(x, t) = \int_0^t k(x, t - y)\varphi(y)\, dy. \tag{1}$$

It is less useful than the h-transform. Accordingly, we shall not discuss it in great detail, restricting ourselves to the degree of generality needed in the remainder of this chapter.

Theorem 11.1

1. $\varphi(y) \in C$, $0 \leq y < \infty$,
2. $u(x, t) = \int_0^t k(x, t - y)\varphi(y)\, dy$

$\Rightarrow$ A. $u(x, t) \in H$, $0 < x < \infty$, $-\infty < t < \infty$,

B. $u(0 +, t) = \dfrac{1}{\sqrt{4\pi}} \int_0^t \dfrac{\varphi(y)}{\sqrt{t - y}}\, dy$, $t > 0$.

As in §8,

$$u_{xx}(x, t) = \int_0^t k_{xx}(x, t - y)\varphi(y)\, dy,$$

$$u_t(x, t) = \int_0^t k_t(x, t - y)\varphi(y)\, dy.$$

Since $k(x, t) \in H$ for $x > 0$, A is proved. To prove B we have

$$\int_0^t k(x, t - y)\varphi(y)\, dy \ll \dfrac{1}{\sqrt{4\pi}} \int_0^t \dfrac{|\varphi(y)|}{\sqrt{t - y}}\, dy.$$

Since the dominant integral is independent of x, we may take the limit under the integral sign to obtain B. Since $u(-x, t) = u(x, t)$, we see that $u(0 -, t) = u(0 +, t) = u(0, t)$, so that $u(x, t) \in C$ for $-\infty < x < \infty$. In contrast, $u(x, t)$ as defined by the h-transform is an odd function of x

and is usually discontinuous at $x = 0$. The function (1) usually does not belong to H in any region including a piece of the t-axis because $u_x(x, t)$ is an h-transform and may have the discontinuity just described.

Theorem 11.2

1. $\varphi(y) \in C$, $\quad 0 \leq y < \infty$,
2. $u(x, t) = \int_0^t k(x, t - y)\varphi(y)\, dy$,
3. $s = \sigma + i\tau$, $\quad t_0 > 0$

$\Rightarrow \quad u(s, t_0) \in A$, $\quad |\arg s| < \pi/4$.

The change of variable $y = t_0 - (4r)^{-1}$ gives

$$u(s, t_0) = \frac{1}{4\sqrt{\pi}} \int_{1/(4t_0)}^{\infty} e^{-s^2 r} \frac{\varphi(t_0 - (1/4r))}{r^{3/2}}\, dr.$$

Now the same argument as that used in §9 gives the desired result.

Corollary 11.2

1. $0 < a < b$, $\quad 0 < t_0$

$\Rightarrow \quad u(x, t_0) \in I$, $\quad a \leq x \leq b$.

This follows also as in §9.

Theorem 11.3

1. $\varphi(y) \in C$, $\quad 0 \leq y < \infty$,
2. $u(x, t) = \int_0^t k(x, t - y)\varphi(y)\, dy$,
3. $0 < c$, $\quad 0 < x_0$

$\Rightarrow \quad u(x_0, t) \in II$, $\quad -c \leq t \leq c$.

We appeal to equation 9(2) and proceed as in the proof of Theorem 9.2. Here we must prove

$$\left| \frac{\partial^n k(x, t)}{\partial x^n} \right| \leq \frac{Mn!}{r^n}, \quad \frac{x_0}{2} \leq x \leq 2x_0, \quad -\infty < t < \infty, \quad (2)$$

where M and r are constants independent of *both* x and t. By Lemma 7, Chapter I,
$$0 \leq k(x, t) \leq \frac{B}{x}, \quad 0 < x < \infty, \quad -\infty < t < \infty.$$
Hence (2) holds for $n = 0$ with $M = 2B/x_0$. For $n = 1, 2, \ldots,$
$$\frac{\partial^n k}{\partial x^n} = -\frac{1}{2} \frac{\partial^{n-1} h}{\partial x^{n-1}},$$
so that we may apply 9(3) to obtain
$$\left| \frac{\partial^n k}{\partial x^n} \right| \leq \frac{1}{2} \frac{M(n-1)!}{r^{n-1}} \leq \frac{(Mr)n!}{2r^n},$$
from which (2) follows after suitable redefinition of the constants.

Now
$$\frac{\partial^n u}{\partial t^n} = \int_0^t \frac{\partial^n k}{\partial t^n}(x, t-y)\varphi(y)\,dy = \int_0^t \frac{\partial^{2n}}{\partial x^{2n}} k(x, t-y)\varphi(y)\,dy,$$
$$\left| \frac{\partial^n u}{\partial t^n}(x_0, t) \right| \leq \frac{M(2n)!}{r^{2n}} \int_0^c |\varphi(y)|\,dy, \quad 0 \leq t \leq c.$$

Thus the defining inequalities of the class II are satisfied for $k(x_0, t)$ on $0 \leq t \leq c$. They are satisfied trivially for $t < 0$, and the theorem is proved.

12 A BASIC INTEGRAL REPRESENTATION

Any temperature function can be given an integral representation valid in some neighborhood of any point where it satisfies the heat equation. The basic result follows:

Theorem 12

1. $u(x, t) \in H, \quad a \leq x \leq b, \quad 0 \leq t \leq c;$
2. $a < x < b, \quad 0 < t < c$

$$\Rightarrow u(x, t) = \int_a^b k(x - y, t) u(y, 0)\,dy$$
$$-\tfrac{1}{2}\int_0^t h(a - x, t - y)u(a, y)\,dy + \tfrac{1}{2}\int_0^t h(b - x, t - y)u(b, y)\,dy$$
$$-\int_0^t k(a - x, t - y)u_x(a, y)\,dy + \int_0^t k(b - x, t - y)u_x(b, y)\,dy.$$

12. A BASIC INTEGRAL REPRESENTATION

Thus $u(x, t)$ is the sum of a Poisson transform, two h-transforms, and two k-transforms. To prove this we appeal to Theorem 4, Chapter I, where

$$R = \{(x, t) \mid a \leqslant x \leqslant b,\ 0 \leqslant t \leqslant t_0\}, \qquad 0 < t_0 < c,$$

$$v(x, t) = k(x - x_0, t_0 + \delta - t), \qquad a < x_0 < b,\ t_0 < t_0 + \delta < c,$$

$$v_x(x, t) = -\tfrac{1}{2} h(x - x_0, t_0 + \delta - t).$$

The double integral in Green's theorem is 0 since $u \in H$ and $v \in H^*$ in R. Hence

$$\int_a^b k(x_0 - y, \delta) u(y, t_0)\, dy = \int_a^b k(x_0 - y, t_0 + \delta) u(y, 0)\, dy$$

$$+ \tfrac{1}{2} \int_0^{t_0} h(b - x_0, t_0 - y + \delta) u(b, y)\, dy$$

$$- \tfrac{1}{2} \int_0^{t_0} h(a - x_0, t_0 - y + \delta) u(a, y)\, dy$$

$$- \int_0^{t_0} k(a - x_0, t_0 - y + \delta) u_x(a, y)\, dy$$

$$+ \int_0^{t_0} k(b - x_0, t_0 - y + \delta) u_x(b, y)\, dy. \tag{1}$$

In all six of these transforms, the function transformed is continuous since $u(x, t) \in H$ on the boundary of R. Now let $\delta \to 0$. By Corollary 5.1, the Poisson integral on the left side of (1) approaches $u(x_0, t_0)$. To obtain the remaining limits we have only to set $\delta = 0$ in the integrands, for by Theorems 8 and 11.1, the transforms surely define continuous functions of δ. Since x_0, t_0 are arbitrary, the theorem is proved.

Under the hypotheses of this theorem we see that $u(x, t)$ is completely determined inside the given rectangle by its values on the base and vertical sides and by those of $u_x(x, t)$ on those sides. However, from physical considerations we expect the temperature of a finite bar, insulated except at its ends, to be completely determined by its initial temperature and by the temperatures maintained at the ends. This expectation is already confirmed by Theorem 2 (uniqueness) of Chapter II. Hence, the use of $u_x(x, t)$ in the integral representation must be redundant. Later we shall obtain another integral representation, avoiding its use. It is sometimes convenient to express this representation as a line integral over the complete boundary Γ of a rectangle

$$R = \{(x, t) \mid a \leqslant x \leqslant b,\ 0 \leqslant t \leqslant c\}$$

Corollary 12

1. $u(x, t) \in H$ in R,
2. (x, t) is interior to R

$$\Rightarrow \quad u(x, t) = \int_\Gamma k(x - \xi, t - \eta) u(\xi, \eta) d\xi$$

$$+ \left[k(x - \xi, t - \eta) u_\xi(\xi, \eta) - \tfrac{1}{2} h(x - \xi, t - \eta) u(\xi, \eta) \right] d\eta,$$

where integration is in the counterclockwise direction.

Since $k(x - \xi, t - \eta) \equiv 0$ for $t < \eta$, this is only a restatement of the theorem. We have taken into account that $k(-x, t) = k(x, t)$ and that $h(-x, t) = -h(x, t)$.

13 ANALYTIC CHARACTER OF EVERY TEMPERATURE FUNCTION

By use of Theorem 12 we may now show that every temperature function $u(x, t)$ is analytic in x and belongs to class II in t. If $u(x, t) \in H$ in an arbitrary domain, we can choose constants a, b, c, δ such that $u(x, t - \delta)$ will satisfy the hypotheses of Theorem 12. Hence, it will be sufficient to prove the following theorem.

Theorem 13

1. $u(x, t) \in H$, $a \leqslant x \leqslant b$, $0 \leqslant t \leqslant c$,
2. $a < x_0 < b$, $0 < t_0 < c$,
3. $0 < \delta < \dfrac{b - a}{2}$, $\delta < c$,

$\Rightarrow \quad u(x, t_0) \in \text{I}$, $a + \delta \leqslant x \leqslant b - \delta$,

$u(x_0, t) \in \text{II}$, $\delta \leqslant t \leqslant c$.

Theorem 12 is here applicable, and we have only to examine the five integrals of the representation separately. Denote them, in the order of appearance, by $J_1, \ldots, J_5$. Both h-transforms have similar properties, as do the two k-transforms. Hence we need consider only $J_1, J_2,$ and J_4. By Theorem 4.1, $J_1(x, t_0)$ is entire and hence $\in \text{I}$ on $-\infty < x < \infty$. By Theorem 4.2, $J_1(x_0, t) \in A$ for $0 < t < \infty$ and hence it belongs to II, a fortiori, on $\delta \leqslant t \leqslant c$.

13. ANALYTIC CHARACTER OF EVERY TEMPERATURE FUNCTION

For the h-transform J_2, we apply Corollary 9.1 to conclude that $J_2(x, t_0) \in \mathrm{I}$ on $a + \delta \leqslant x \leqslant b - \delta$. By Theorem 9.2, $J_2(x_0, t) \in \mathrm{II}$ on $\delta \leqslant t \leqslant c$.

For the k-transform J_4, we obtain the desired conclusions by use of Corollary 11.2 and Theorem 11.3. This concludes the proof.

It is noteworthy that, although continuity of only the early derivatives of $u(x, t)$ is postulated in the definition of H, the additional demand that two of the derivatives should be equal, $u_{xx} = u_t$, automatically brings with it C^∞ for $u(x, t)$ and even analyticity in the space variable.

Observe that the proof of Theorem 4, Chapter II, is now complete. There we needed to know the existence and continuity of some of the higher derivatives of a temperature function.

Chapter V

THETA-FUNCTIONS

1 INTRODUCTION

In Chapter II we introduced the boundary-value problem for a finite bar and later, Theorem 5, solved it in the special case when the ends of the bar are held at zero temperature. To solve the problem in the general case it is convenient to use one of the theta-functions of C. G. J. Jacobi. In §5 of Chapter I we introduced two of these functions for which we used the notation $\theta(x, t)$ and $\varphi(x, t)$. We observed that they satisfy the heat equation and that they are periodic in x. It is this periodicity that makes them useful in the problem of the finite bar, as we shall see.

Definition 1a[†]

For $(x, t) \neq (2m\pi, 0)$, $m = 0, \pm 1, \pm 2, \ldots,$

$$\theta(x, t) = \sum_{n=-\infty}^{\infty} k(x + 2n\pi, t). \tag{1}$$

Definition 1b

For $(x, t) \neq (2m\pi, 0)$, $m = 0, \pm 1, \pm 2, \ldots,$

$$\varphi(x, t) = \sum_{n=-\infty}^{\infty} h(x + 2n\pi, t). \tag{2}$$

[†] In Jacobi's notation this function is $(2\pi)^{-1}\theta_3(x/2, e^{-t})$.

1. INTRODUCTION

To study the convergence of these series it will be useful to have upper bounds on $k(x, t)$ and $h(x, t)$ and their derivatives. Following our convention introduced earlier, primes and superscripts refer to differentiation with respect to x. In the following lemma, the letter A stands for a constant which may change from function to function.

Lemma 1

$\Rightarrow$

1. $p > -\frac{1}{2}$, $\quad t > 0, \quad x > 0$

   A. $k(x, t) \leqslant \dfrac{At^p}{x^{2p+1}}$,

   B. $|k'(x, t)| \leqslant \dfrac{At^p}{x^{2p+2}}$,

   C. $|k''(x, t)| \leqslant \dfrac{At^p}{x^{2p+3}}$.

To prove the first inequality write

$$k(x, t) = \frac{t^p e^{-x^2/(4t)}}{\sqrt{4\pi}\, t^{p+(1/2)}},$$

and apply Lemma 7 of Chapter I with $y = 1/t$, $c = x^2/4$, p replaced by $p + \frac{1}{2}$. This yields the desired upper bound with the constant

$$A = \frac{1}{\sqrt{4\pi}} \left(\frac{4p + 2}{e} \right)^{(2p+1)/2}.$$

Conclusions B and C follow easily from the equations

$$k' = -\left(\frac{x}{2t} \right) k, \qquad k'' = \frac{k(x^2 - 2t)}{4t^2}.$$

Since $h(x, t) = (x/t)k(x, t)$, the lemma may be used in an obvious way to obtain bounds on the derivatives of $h(x, t)$. It may be used for negative x since $k(-x, t) = k(x, t)$.

Theorem 1.1 $\theta(x, t)$ and $\varphi(x, t)$ are well-defined for all (x, t) except $(2m\pi, 0)$, $m = 0, \pm 1, \pm 2, \ldots$.

We recall our convention that $k(x, t)$ and $h(x, t)$ are identically 0 for $t < 0$, so there is nothing to prove there. For $t > 0$, apply Lemma 1 with $p = 1$:

$$\sum_{n=-\infty}^{\infty} k(x + 2n\pi, t) \ll \sum_{-\infty}^{\infty} \frac{At}{|x + 2n\pi|^3}. \qquad (3)$$

The dominant series, and hence series (1), converges if $x \neq 2m\pi$, $m = 0$, $\pm 1, \ldots$. Hence series (1) converges in the half-plane $t > 0$. That it converges for $t = 0$, away from the exceptional points, is obvious. One term of the series is undefined at each of those points. Since $h(x, t) = (x/t)k(x, t)$, a dominant series for series (2) is

$$\sum_{n=-\infty}^{\infty} \frac{A}{(x + 2n\pi)^2} < \infty,$$

and the theorem is proved.

Theorem 1.2

$$\theta(x + 2\pi, t) = \theta(x, t), \quad t > 0;$$

$$\varphi(x + 2\pi, t) = \varphi(x, t), \quad t > 0.$$

If $m = n + 1$,

$$\sum_{n=-\infty}^{\infty} k(x + 2n\pi + 2\pi, t) = \sum_{m=-\infty}^{\infty} k(x + 2m\pi, t).$$

Replacing k by h in this equation gives the periodicity of φ, as stated.

2 ANALYTICITY

The θ-functions are analytic in each variable separately as follows:

Theorem 2.1

1. $s = \sigma + i\tau$, $\quad 0 < t_0$, $\quad p = 0, 1, 2, \ldots$

$\Rightarrow$ A. $\theta(s, t_0) \in A$ (analytic), $\quad |s| < \infty$,

B. $\varphi(s, t_0) \in A$, $\quad |s| < \infty$,

C. $\dfrac{\partial^p \theta(k, t)}{\partial x^p} = \sum_{n=-\infty}^{\infty} \dfrac{\partial^p k(x + 2n\pi, t)}{\partial x^p}, \quad -\infty < x < \infty,$

D. $\dfrac{\partial^p \varphi(x, t)}{\partial x^p} = \sum_{n=-\infty}^{\infty} \dfrac{\partial^p h(x + 2n\pi, t)}{\partial x^p}, \quad -\infty < x < \infty.$

For an arbitrary number $R > 0$, we have for $|s| \leqslant R$

$$\sum_{|n| > R/(2\pi)} e^{-(s + 2n\pi)^2/(4t_0)} \ll e^{R^2/(4t_0)} \sum_{|n| > R/(2\pi)} e^{-(2\pi|n| - R)^2/(4t_0)} < \infty.$$

Since the dominant series is independent of s, the series

$$\theta(s, t_0) = \sum_{n=-\infty}^{\infty} k(s + 2n\pi, t_0)$$

converges uniformly for $|s| \leq R$. Hence by Weierstrass's theorem, $\theta(s, t_0) \in A$ for $|s| < R$ and term-by-term differentiation with respect to s is permissible. Hence, A and C are proved. Since $h(s, t) = -2k'(s, t)$, conclusions B and D also follow, and $\varphi(s, t) = -2\theta'(s, t)$.

Theorem 2.2

1. $s = \sigma + i\tau$, $\quad -\infty < x_0 < \infty, \quad p = 0, 1, 2, \ldots$

$\Rightarrow$
- A. $\theta(x_0, s) \in A,$ $\hfill \sigma > 0,$
- B. $\varphi(x_0, s) \in A,$ $\hfill \sigma > 0,$
- C. $\dfrac{\partial^p \theta(x, t)}{\partial t^p} = \sum_{n=-\infty}^{\infty} \dfrac{\partial^p k(x + 2n\pi, t)}{\partial t^p}, \quad t > 0,$
- D. $\dfrac{\partial^p \varphi(x, t)}{\partial t^p} = \sum_{n=-\infty}^{\infty} \dfrac{\partial^p h(x + 2n\pi, t)}{\partial t^p}, \quad t > 0.$

For an arbitrary number $R > 0$, we have for Re $1/s \geq 1/R$

$$\sum_{n=-\infty}^{\infty} e^{-(x_0+2n\pi)^2/(4s)} \ll \sum_{n=-\infty}^{\infty} e^{-(x_0+2n\pi)^2/(4R)} < \infty, \tag{1}$$

and again we may apply the Weierstrass theorem to conclude that the series (1) converges uniformly in the region

$$\text{Re } \frac{1}{s} = \frac{\sigma}{\sigma^2 + \tau^2} \leq \frac{1}{R}.$$

This is a disk with center at $(R/2, 0)$ and radius $R/2$. Since such a disk can be made to cover any point in the half-plane $\sigma > 0$, the proof may be concluded as for the previous theorem.

3 θ-FUNCTIONS IN H

Since the general terms of series 1(1) and 1(2) are temperature functions, the same should be true of their sums.

Theorem 3

$$1. \quad (x, t) \neq (2m\pi, 0), \quad m = 0, \pm 1, \pm 2, \ldots$$

$$\Rightarrow \quad \theta(x, t) \in H, \quad \varphi(x, t) \in H.$$

For $t > 0$, the result is an obvious consequence of Theorems 2.1 and 2.2 because term-by-term differentiation is valid and each term of both series $\in H$ when $t > 0$. Since $\theta(x, t) = \varphi(x, t) = 0$ for $t < 0$, there is nothing to prove there. We need only consider the situation on the x-axis. By Lemma 1

$$\sum_{n=-\infty}^{\infty} k''(x + 2n\pi, t) \ll \sum_{n=-\infty}^{\infty} \frac{A}{|x + 2n\pi|^3}. \tag{1}$$

The dominant series, independent of t, converges when $x \neq 2m\pi$ and $m = 0, \pm 1, \ldots$. Hence series (1) converges uniformly and consequently represents a continuous function of t for $-\infty < t < \infty$, $x \neq \pm 2m\pi$. Therefore

$$\sum_{n=-\infty}^{\infty} k_{xx}(x + 2n\pi, 0+) = \sum_{n=-\infty}^{\infty} k_t(x + 2n\pi, 0+) = 0$$

and $\theta(x, t)$ also $\in H$ on the x-axis away from the exceptional points. A similar proof holds for $\varphi(x, t)$.

4 ALTERNATE EXPANSIONS

Considered as functions of x, the functions $\theta(x, t)$ and $\varphi(x, t)$ have expansions in Fourier series.

Theorem 4.1

$$1. \quad 0 < t, \quad -\infty < x < \infty$$

$$\Rightarrow \quad \theta(x, t) = \frac{1}{2\pi} \sum_{n=-\infty}^{\infty} e^{nix - n^2 t} \tag{1}$$

$$= \frac{1}{2\pi} + \frac{1}{\pi} \sum_{n=1}^{\infty} e^{-n^2 t} \cos nx. \tag{2}$$

Since $\theta(x, t)$ has period 2π and surely satisfies the classic Dirichlet conditions for expansibility in Fourier series, by Theorem 2.1, it is only

4. ALTERNATE EXPANSIONS

necessary to compute the Fourier coefficients. For the exponential form (1) they are

$$a_m(t) = \frac{1}{2\pi} \int_0^{2\pi} \theta(x, t) e^{-mix} dx, \quad m = 0, \pm 1, \ldots. \quad (3)$$

In §2 we showed that the series defining $\theta(x, t)$ converges uniformly in $|x| \leq R$ for every R. Hence we may substitute that series in (3) and integrate term by term,

$$a_m(t) = \frac{1}{2\pi} \int_0^{2\pi} \sum_{n=-\infty}^{\infty} e^{-imx} k(x + 2n\pi, t) \, dx$$

$$= \frac{1}{2\pi} \sum_{n=-\infty}^{\infty} \int_{2n\pi}^{(2n+2)\pi} e^{-imx} k(x, t) \, dx$$

$$= \frac{1}{2\pi} \int_{-\infty}^{\infty} e^{-imx} k(x, t) \, dx.$$

Here we have replaced $x + 2n\pi$ by x, used the periodicity of e^{-imx} and then have collapsed the sum of integrals into a single integral. The latter integral is a bilateral Laplace transform, which we evaluated in Theorem 7.1, Chapter III. It has the value $e^{-m^2 t}$, so that equation (1) is valid. Equation (2) follows from (1) by grouping terms in an obvious way.

Corollary 4.1

1. $0 < t, \quad -\infty < x < \infty, \quad -\infty < y < \infty$

$$\Rightarrow \quad \theta(x - y, t) - \theta(x + y, t) = \frac{2}{\pi} \sum_{n=1}^{\infty} e^{-n^2 t} \sin nx \sin ny.$$

This combination of θ-functions will be useful later. Its series expansion follows from (2).

Theorem 4.2

1. $0 < t < \infty, \quad -\infty < x < \infty$

$$\Rightarrow \quad \varphi(x, t) = -\frac{i}{\pi} \sum_{n=-\infty}^{\infty} n e^{nix - n^2 t} \quad (4)$$

$$= \frac{2}{\pi} \sum_{n=1}^{\infty} n e^{-n^2 t} \sin nx. \quad (5)$$

Since $\varphi(x, t) = -2\theta'(x, t)$, the present expansions follow at once by differentiating equations (1) and (2). The step is valid since series (4) and (5) converge uniformly on $-\infty < x < \infty$ for a fixed $t > 0$.

5 TWO POSITIVE KERNELS

We shall presently be considering two new integral transforms, the kernels of which are involved with θ-functions. One of the kernels is the function of Corollary 4.1, the other is $\varphi(x, t - y)$. We shall need the fact that these kernels are ≥ 0 in certain regions of the x, t-plane. The fact becomes more transparent if one uses an infinite product expansion for $\theta(x, t)$ derived by Jacobi and recorded, for example, by Whittaker and Watson [1943; 469]:

$$\theta(x, t) = C(t) \prod_{n=1}^{\infty} (1 + 2q^{2n-1} \cos x + q^{4n-2}), \qquad q = e^{-t}; \quad (1)$$

$$C(t) = \frac{1}{2\pi} \prod_{n=1}^{\infty} (1 - q^{2n}).$$

Theorem 5.1

1. $0 < x < \pi$, $\quad 0 < y < \pi$, $\quad 0 < t$

$\Rightarrow \qquad \theta(x - y, t) - \theta(x + y, t) > 0.$

Observe that each factor of the infinite product (1) is positive for all x and for all $t > 0$:

$$(1 + 2q^{2n-1} \cos x + q^{4n-2}) = (1 + q^{2n-1} \cos x)^2 + q^{4n-2} \sin^2 x. \quad (2)$$

Hence, to prove the theorem we need only compare the general factors corresponding to $\theta(x - y, t)$ and $\theta(x + y, t)$. We have

$$1 + 2q^{2n-1} \cos(x - y) + q^{4n-2} > 1 + 2q^{2n-1} \cos(x + y) + q^{4n-2}$$

since

$$\cos(x - y) - \cos(x + y) = 2 \sin x \sin y > 0.$$

Theorem 5.2

1. $0 < x < \pi$, $\quad 0 < t$

$\Rightarrow \qquad \varphi(x, t) > 0.$

5. TWO POSITIVE KERNELS

From the equation $\varphi(x, t) = -2\theta'(x, t)$, we have by logarithmic differentiation of equation (1) that

$$\frac{\varphi(x, t)}{4\theta(x, t)} = \sin x \sum_{n=1}^{\infty} \frac{q^{2n-1}}{1 + 2q^{2n-1}\cos x + q^{4n-2}}. \tag{3}$$

This step is justified by the general Weierstrass theory of infinite product expansions. However, one may verify directly that the series (3) is dominated by the convergent series

$$\sum_{n=1}^{\infty} \frac{q^{2n-1}}{(1 - q^{2n-1})^2}$$

and hence converges. Every term of series (3) is positive by (2). Since $\theta(x, t) > 0$ for all x when $t > 0$, by Definition 1a, the conclusion now follows from equation (3).

We now give alternate proofs of the previous theorems which will make no appeal to the infinite product (1). The tool is rather the maximum principle for the heat equation. First we prove Theorem 5.2.

Apply Corollary 3.1, Chapter II, to the function $\varphi(x, t)$ for the rectangle

$$R = \{(x, t) \mid 0 \leq x \leq \pi, \ 0 \leq t \leq c\}.$$

From equation 4(5) we see that $\varphi(x, t) \in C$ and $= 0$ on the vertical sides of R except at $(0, 0)$. On the base we show that

$$\lim_{(x, t) \to (x, 0+)} \varphi(x, t) \geq 0, \qquad 0 \leq x \leq \pi. \tag{4}$$

By dropping the term $h(x, t)$ which is ≥ 0 in R from the series 1(2), we have

$$\varphi(x, t) \geq \sum_{|n|>0} h(x + 2n\pi, t), \qquad (x, t) \in R. \tag{5}$$

By Lemma 1 with $p = 2$,

$$\sum_{|n|>0} |h(x + 2n\pi, t)| \ll \sum_{|n|>0} \frac{At}{(x + 2n\pi)^4}$$

$$\ll \sum_{|n|>0} \frac{At}{(2|n|\pi - \pi)^4}, \qquad (x, t) \in R. \tag{6}$$

Since the right-hand side of (6) $\to 0$ when $t \to 0$, we see that (4) follows from (5) and (6). The hypotheses of Corollary 3.1, Chapter II, are satisfied, so that $\varphi(x, t) \geq 0$ in R, as desired.

To prove Theorem 5.1, observe first that

$$2 \sin nx \sin ny = \cos n(x - y) - \cos n(x + y)$$
$$= \cos n(x - y) - \cos n(2\pi - x - y)$$
$$= n \int_{x-y}^{x+y} \sin nr\, dr = n \int_{x-y}^{2\pi - x - y} \sin nr\, dr. \tag{7}$$

These equations combined with Corollary 4.1 and Theorem 4.2 show that

$$\theta(x - y, t) - \theta(x + y, t) = \tfrac{1}{2} \int_{x-y}^{x+y} \varphi(r, t)\, dr = \tfrac{1}{2} \int_{x-y}^{2\pi - x - y} \varphi(r, t)\, dr. \tag{8}$$

The term-by-term integration is valid since for fixed $t > 0$ series 4(4) is uniformly convergent for $-\infty < x < \infty$. Suppose, without loss of generality, that $x \geqslant y$. Then the interval of integration of one or the other of the integrals (8) is included in the interval $(0, \pi)$, where the integrand is $\geqslant 0$. The same must be true of the left-hand side, as we wished to prove. I am indebted to R. A. Askey for calling my attention to the usefulness of identity (7).

6 A θ-TRANSFORM

Here we define an integral transform using the kernel of Theorem 5.1:

$$u(x, t) = \int_0^\pi [\theta(x - y, t) - \theta(x + y, t)] f(y)\, dy. \tag{1}$$

As usual, $f(y) \in L$ on $(0, \pi)$ without further statement. Defined initially on $(0, \pi)$ only, we may extend its definition to $-\infty < y < \infty$ by demanding that it should be odd and should have period 2π. Transform (1) then becomes equivalent to the Poisson transform. We treat the more general Stieltjes integral, in which case the integrator function

$$\alpha(y) = \int_0^y f(r)\, dr$$

must be even.

Theorem 6.1

1. $u(x, t) = \int_0^\pi [\theta(x - y, t) - \theta(x + y, t)]\, d\alpha(y),$ (2)
2. $\alpha(-y) = \alpha(y),$ $\qquad -\infty < y < \infty,$
3. $\alpha(y + 2\pi) = \alpha(y),$ $\qquad -\infty < y < \infty,$

$\Rightarrow \qquad u(x, t) = \int_{-\infty}^\infty k(x - y, t)\, d\alpha(y).$

Observe first that the integral (2) surely exists for all x since $\theta(x, t) \in A$, $-\infty < x < \infty$, and $\alpha(y) \in V$, $0 \leq y \leq \pi$. By 2 and 3

$$\int_{-\pi}^{0} \theta(x - y, t) \, d\alpha(y) = -\int_{0}^{\pi} \theta(x + y, t) \, d\alpha(y),$$

$$u(x, t) = \int_{-\pi}^{\pi} \theta(x - y, t) \, d\alpha(y),$$

$$\int_{(2n-1)\pi}^{(2n+1)\pi} k(x - y, t) \, d\alpha(y) = \int_{0}^{\pi} k(x - y - 2n\pi, t) \, d\alpha(y).$$

Hence,

$$\int_{-\infty}^{\infty} k(x - y, t) \, d\alpha(y) = \sum_{n=-\infty}^{\infty} \int_{(2n-1)\pi}^{(2n+1)\pi} k(x - y, t) \, d\alpha(y)$$

$$= \sum_{n=-\infty}^{\infty} \int_{-\pi}^{\pi} k(x - y - 2n\pi, t) \, d\alpha(y)$$

$$= \int_{-\pi}^{\pi} \sum_{n=-\infty}^{\infty} k(x - y - 2n\pi, t) \, d\alpha(y)$$

$$= \int_{-\pi}^{\pi} \theta(x - y, t) \, d\alpha(y) = u(x, t).$$

The series defining the θ-function was shown to converge uniformly in every finite interval in §2, so that the term-by-term integration over $(-\pi, \pi)$, which we just did, was valid. This completes the proof.

Corollary 6.1

$$u(x, t) \in H, \quad 0 < t, \quad -\infty < x < \infty.$$

This follows by Theorem 3, Chapter IV.

Theorem 6.2

1. $u(x, t) = \int_{0}^{\pi} [\theta(x - y, t) - \theta(x + y, t)] f(y) \, dy$ \quad (3)

$\Rightarrow$ A. $\quad u(x, 0+) = f(x), \quad$ almost all x, $\quad 0 < x < \pi$,

B. $\quad u(0, t) = u(\pi, t) = 0, \quad 0 < t.$

Conclusion A follows from the previous theorem and our knowledge of the Poisson transform, Corollary 5.2 of Chapter IV. Since $\theta(x, t)$ is even in

x, $\theta(-y, t) - \theta(y, t) = 0$ and $\theta(\pi - y, t) - \theta(\pi + y, t) = \theta(\pi - y, t) - \theta(-\pi - y, t)$. The latter difference is zero by periodicity, and B is proved.

Corollary 6.2a

$$2. \quad f(a+), f(a-) \text{ exist}, \quad 0 < a < \pi,$$

$\Rightarrow$
$$u(x, 0+) = \frac{f(a+) + f(a-)}{2}.$$

This follows from Corollary 5.2, Chapter IV.

Corollary 6.2b

$$2. \quad f(y) \in C, \quad 0 \leqslant y \leqslant \pi,$$
$$3. \quad f(0) = f(\pi) = 0$$

$\Rightarrow$
$$\lim_{t \to 0+} u(x, t) = f(x), \quad \text{uniformly on } 0 \leqslant x \leqslant \pi.$$

Hypothesis 3 guarantees that the definition of $f(y)$ can be extended so as to make $f(y)$ periodic, odd and continuous on $-\infty < y < \infty$. Thus Theorem 6.1 is applicable and we may appeal to Corollary 1, Chapter IV for our conclusion.

This corollary enables us to give a more general solution of the boundary-value problem treated in Theorem 5, Chapter II. There we demanded that the initial function $f(x)$ should belong to C^2 on $0 \leqslant x \leqslant \pi$ in order to be assured that the Fourier series there used should converge to $f(x)$. Now we demand simple continuity, and Corollary 6.2b assures us that the function (3) provides the unique solution of the given problem, even though the Fourier series for the initial function may diverge. See Titchmarsh [1939; 416].

Theorem 6.3

$$1. \quad u(x, t) = \int_0^\pi [\theta(x - y, t) - \theta(x + y, t)] \, d\alpha(y),$$
$$2. \quad 0 < a < b < \pi$$

$\Rightarrow$

A. $\displaystyle\lim_{t \to 0+} \int_a^b u(x, t) \, dx = \frac{\alpha(b+) + \alpha(b-)}{2} - \frac{\alpha(a+) + \alpha(a-)}{2}$

B. $\displaystyle\lim_{t \to 0+} \int_0^\pi u(x, t) \, dx = \alpha(\pi-) - \alpha(0+).$

This is a consequence of Theorem 6.1 and the general inversion theorem of Chapter IV, Theorem 6. Observe that under hypotheses 2 and 3 of Theorem 6.1, $\alpha(0-) = \alpha(0+)$ and $\alpha(\pi+) = \alpha(\pi-)$.

7 A φ-TRANSFORM

We use next the kernel $\varphi(x,t)$ to define the φ-transform of a function $f(y)$ as

$$u(x, t) = \int_0^t \varphi(x, t-y) f(y)\, dy. \tag{1}$$

We show first that it also defines a temperature function in the vertical strip $0 < x < 2\pi$, and of course in congruent strips, by periodicity. We treat the more general Stieltjes integral.

Theorem 7.1

$$1. \quad u(x, t) = \int_0^t \varphi(x, t-y)\, d\alpha(y)$$

$$\Rightarrow \quad u(x, t) \in H, \quad 0 < x < 2\pi, \quad -\infty < t < \infty.$$

For a fixed positive t we have by Lemma 1

$$\sum_{n=-\infty}^{\infty} h(x + 2n\pi, t - y) \ll \sum_{n=-\infty}^{\infty} \frac{A}{(x+2n\pi)^2} < \infty, \quad 0 < x < 2\pi.$$

Hence the series on the left converges uniformly for $0 \le y \le t$, and we may integrate term-by-term to obtain

$$u(x, t) = \sum_{n=-\infty}^{\infty} \int_0^t h(x + 2n\pi, t - y)\, d\alpha(y). \tag{2}$$

The general term of this series $\in H$ for $x \ne 2n\pi$ by Theorem 8, Chapter IV, $(h(-x, t) = -h(x, t))$. If term-by-term differentiation is valid, the desired conclusion will follow. We know from Theorem 8, Chapter IV, that the integral (1) may be differentiated with respect to x or t under the integral sign. Hence it will be sufficient to show that

$$\sum_{|n|\ge 2}^{\infty} \int_0^t h''(x + 2n\pi, t-y)\, d\alpha(y) \tag{3}$$

converges uniformly for $|x| \leq 2\pi$ and $-c \leq t \leq c$, where c is arbitrary. If the variation of $\alpha(y)$ on $0 \leq y \leq c$ is V, we have by Lemma 1 that series (3) is dominated by

$$\sum_{|n|>2} \frac{AV}{(x+2n\pi)^4} \ll \sum_{|n|>2} \frac{AV}{(2n\pi - 2\pi)^4} < \infty, \qquad |x| \leq 2\pi, \ |t| \leq c.$$

Since the final series is independent of x and t, and since c is arbitrary, the theorem is proved.

Corollary 7.1

A. $\quad u(\pi -, t) = 0, \hspace{4cm} 0 \leq t < \infty;$

B. $\quad u(x, 0+) = 0, \hspace{4cm} 0 \leq x \leq \pi;$

C. $\quad u(x, t) - \int_0^t h(x, t-y)\, d\alpha(y) = 0$

and $\in C$ on $t = 0, 0 \leq x \leq \pi$ and on $x = 0, 0 \leq t < \infty$.

For A and B we may write

$$u(x, t) = \sum_{n=0}^{\infty} \int_0^t [h(x + 2n\pi, t-y)$$

$$+ h(x - 2n\pi - 2\pi, t-y)]\, d\alpha(y),$$

$$u(\pi, t) = \sum_{n=0}^{\infty} \int_0^t [h(2n\pi + \pi, t-y) + h(-2n\pi - \pi, t-y)]\, d\alpha(y). \tag{4}$$

Since $h(-x, t) = -h(x, t)$, we have $u(\pi, t) = 0$. Also, by Theorem 7.1, $u(x, t)$ is surely continuous on the line $x = \pi$ so that $u(\pi -, t) = u(\pi, t)$. Similarly for B. Observe that $u(0, 0+)$ is 0 even though the first term of series (4) may not belong to H at $(0, 0)$.

For C we have

$$u(x, t) - \int_0^t h(x, t-y)\, d\alpha(y)$$

$$= \sum_{n=1}^{\infty} \int_0^t [h(x + 2n\pi, t-y) + h(x - 2n\pi, t-y)]\, d\alpha(y). \tag{5}$$

Each term of the series vanishes for $x = 0$ and its sum $\in H$ there. The desired conclusions are thus evident. We turn next to the inversion of the integral (1).

Theorem 7.2

$$1. \quad u(x, t) = \int_0^t \varphi(x, t - y) f(y) \, dy$$

$\Rightarrow$ A. $u(0 +, t) = f(t)$ almost everywhere.

Isolating one term of the series defining $u(x, t)$, as in equation (5), we have only to apply C of Corollary 7.1 and Corollary 10.1a of Chapter IV to obtain this result.

Corollary 7.2

$$2. \quad f(t_0 -) = f(t_0), \qquad t_0 > 0,$$

$\Rightarrow$ $u(0 +, t_0) = f(t_0).$

This follows in the same way by use of Corollary 10.1b, Chapter IV.

For our purposes, it is important to know that if the given function $f(y) \in C$ on $0 \leq y \leq c$ with $f(0) = 0$, then the function (1) is continuous in the rectangle

$$R = \{(x, t) \mid 0 \leq x \leq \pi, \ 0 \leq t \leq c\},$$

assuming that it has been defined as $f(t)$, its limiting value, on the t-axis.

Theorem 7.3

1. $u(x, t) = \int_0^t \varphi(x, t - y) f(y) \, dy;$
2. $f(y) \in C,$ $0 \leq y \leq c;$
3. $f(0) = 0$

$\Rightarrow$ $u(x, t) \in C$ in R.

By Corollary 7.3, Chapter IV, the function

$$\int_0^t h(x, t - y) f(y) \, dy$$

is continuous in R. As in the proof of Corollary 7.1, the sum of the series

$$\sum_{n=1}^{\infty} \int_0^t [h(x + 2n\pi, t) + h(x - 2n\pi, t)] f(y) \, dy$$

belongs to H in R and hence is surely continuous there. This completes the proof.

8 FOURIER'S RING

A circular ring of fine wire has its initial temperature given as a function $f(x)$ of its arc length x. It is required to determine its subsequent temperatures. Fourier [1878; 265] solved it, without exact assumptions about $f(x)$, by use of trigonometric series. In rearranging his solution he came upon series 4(2) whose sum is $\theta(x, t)$. This may have been the first use of one of the theta-functions studied later in detail by Jacobi.

By imposing Dirichlet's or Jordan's conditions, say, on $f(x)$, we could easily follow Fourier's solution rigorously. The Fourier series involved would then converge and would provide the solution. If continuity alone is assumed that series might diverge. We bypass the Fourier series entirely and give the solution in terms of $\theta(x, t)$, assuming only continuity. Moreover, with these data we produce necessary and sufficient conditions.

We may assume the radius of the ring to be unity. Furthermore we may extend the wire in a straight line along the x-axis from 0 to 2π and assume periodicity in $f(x)$. This equivalent boundary-value problem reads: Find $u(x, t)$ such that

$$u(x, t) \in H, \qquad 0 \leqslant x \leqslant 2\pi, \; 0 < t < \infty,$$
$$u(x, 0+) = f(x) \in C, \qquad 0 \leqslant x \leqslant 2\pi,$$
$$f(0) = f(2\pi).$$

We show that the problem has a unique solution provided by a theta-transform. The result follows.

Theorem 8 Let $f(x) \in C$ on $0 \leqslant x \leqslant 2\pi$, $f(0) = f(2\pi)$. Then

1. $u(x, t) \in H$, $\qquad 0 \leqslant x \leqslant 2\pi, \; 0 < t < \infty,$
2. $\lim_{t \to 0} u(x, t) = f(x) \qquad$ uniformly on $\; 0 \leqslant x \leqslant 2\pi$

$\Leftrightarrow$

$$u(x, t) = \int_0^{2\pi} \theta(x - y, t) f(y) \, dy. \qquad (1)$$

If conditions 1 and 2 hold, and if we define $u(x, 0) = f(x)$, then $u(x, t) \in C$ in the rectangle

$$R = \{(x, t) \mid 0 \leqslant x \leqslant 2\pi, \; 0 \leqslant t \leqslant c\}, \qquad \text{any } c > 0.$$

Hence we may apply Theorem 3.2, Chapter II, to be assured that there is at most one function satisfying 1 and 2. We need only show that the

9. A SOLUTION OF THE FIRST BOUNDARY-VALUE PROBLEM

function (1) satisfies those conditions.

Define $f(x)$ outside the interval $(0, 2\pi)$ so as to have period 2π. Since $f(0) = f(2\pi)$, it will then be continuous on $(-\infty, \infty)$. As in the proof of Theorem 6.1, we have

$$\int_{-\infty}^{\infty} k(x-y, t)f(y)\,dy = \sum_{n=-\infty}^{\infty} \int_{2n\pi}^{(2n+2)\pi} k(x-y, t)f(y)\,dy$$

$$= \int_0^{2\pi} f(y) \sum_{n=-\infty}^{\infty} k(x-y-2n\pi, t)\,dy$$

$$= u(x, t). \tag{2}$$

By Corollary 1 and Theorem 3, Chapter IV, the Poisson transform (2) satisfies 1 and 2. This completes the proof.

Observe that the theorem would be false without the uniformity of approach required by hypothesis 2, for then the solution of the problem would not be unique. The function $\varphi(x, t)$ could be added to the function (1) to produce a second solution.

9 A SOLUTION OF THE FIRST BOUNDARY-VALUE PROBLEM

The θ-transform studied in this chapter enables us to give a solution of Problem I, as posed in Chapter II, §1, with very general boundary conditions. In fact, we assume here only that the given boundary functions are Lebesgue integrable.

Theorem 9

1. $f(x), g_1(x), g_2(x) \in L$ in finite intervals,
2. $u(x, t) = \int_0^{\pi} [\theta(x-y, t) - \theta(x+y, t)]f(y)\,dy$

$$+ \int_0^t \varphi(x, t-y)g_1(y)\,dy$$

$$+ \int_0^t \varphi(\pi-x, t-y)g_2(y)\,dy \tag{1}$$

$\Rightarrow$ A. $u(x, t) \in H$, $0 < x < \pi$, $0 < t < \infty$,

      B. $u(x, 0+) = f(x)$, $0 < x < \infty$,   a.e.,

      C. $u(0+, t) = g_1(t)$, $0 < t < \infty$,   a.e.,

      D. $u(0, \pi-) = g_2(t)$, $0 < t < \infty$,   a.e. .

102   V. THETA-FUNCTIONS

This theorem is an immediate consequence of Theorems 6.2 and 7.2. Each of the integrals (1) has a nonzero limit for just one of the three boundaries in question. To treat the last of these integrals by Theorem 7.2, replace x by $\pi - x$. This transformation obviously carries H into H.

10 UNIQUENESS

The solution to Problem I exhibited in the previous section is in general not unique. If the problem is restated, as in Chapter II, §2, it can now be solved uniquely by θ-transforms. We need two preliminary results.

Lemma 10a

1. $0 < x < \pi, \quad 0 < t$

$$\Rightarrow \quad 1 = \int_0^\pi [\theta(x - y, t) - \theta(x + y, t)]\, dy$$

$$+ \int_0^t \varphi(x, t - y)\, dy + \int_0^t \varphi(\pi - x, t - y)\, dy.$$

From Theorem 9 it is clear that the sum of these three integrals $\in H$ in $0 < x < \pi$ and $0 < t < \infty$ and approaches 1 as (x, t) approaches boundaries as indicated in B–D. The function 1 has the same properties, but we cannot therefore assume the equation of the lemma. As noted earlier, we cannot assume uniqueness when the boundary approach is along the normal, as in B–D. We draw our conclusion by performing the integration explicitly. We use the expansion of Theorem 4.2:

$$\theta(x - y, t) - \theta(x + y, t) = \frac{2}{\pi} \sum_{n=1}^\infty e^{-n^2 t} \sin nx \sin ny, \quad (1)$$

$$\varphi(x, t) = \frac{2}{\pi} \sum_{n=1}^\infty n e^{-n^2 t} \sin nx. \quad (2)$$

For fixed x and t, $t > 0$, series (1) clearly converges uniformly for $-\infty < y < \infty$, so that term-by-term integration is valid, producing

$$\int_0^\pi [\theta(x - y, t) - \theta(x + y, t)]\, dy = \frac{2}{\pi} \sum_{n=1}^\infty \frac{1 - \cos n\pi}{n} e^{-n^2 t} \sin nx.$$
$$(3)$$

10. UNIQUENESS

Also

$$\int_0^t \varphi(x, t - y) \, dy = \int_0^t \varphi(x, y) \, dy = \frac{2}{\pi} \sum_{n=1}^{\infty} \frac{1 - e^{-n^2 t}}{n} \sin nx; \quad (4)$$

$$\int_0^t \varphi(\pi - x, y) \, dy = -\frac{2}{\pi} \sum_{n=1}^{\infty} \frac{1 - e^{-n^2 t}}{n} \cos n\pi \sin nx. \quad (5)$$

Deferring discussion of validity of equations (4) and (5), we add the three integrals (3)–(5) to obtain

$$\frac{2}{\pi} \sum_{n=1}^{\infty} \frac{1 - \cos n\pi}{n} \sin nx = \frac{4}{\pi} \sum_{n=1}^{\infty} \frac{\sin(2n - 1)x}{(2n - 1)}.$$

This is the familiar Fourier series [Widder, 1961c; 394], whose sum is 1 for $0 < x < \pi$.

To justify equation (4) observe first that for fixed x, series (2) converges uniformly in $\epsilon \leqslant t < \infty, \epsilon > 0$. This follows from

$$\sum_{n=1}^{\infty} ne^{-n^2 t} \sin nx \ll \sum_{n=1}^{\infty} ne^{-n^2 \epsilon} < \infty.$$

Hence

$$\int_\epsilon^t \varphi(x, y) \, dy = \frac{2}{\pi} \sum_{n=1}^{\infty} \frac{e^{-\epsilon n^2} - e^{-n^2 t}}{n} \sin nx.$$

The series

$$\sum_{n=1}^{\infty} e^{-\epsilon n^2} \frac{\sin nx}{n}$$

is a Dirichlet series in ϵ which converges for $\epsilon = 0$. Hence it converges uniformly for $0 \leqslant \epsilon < \infty$ [Widder, 1971; 27]. Hence

$$\lim_{\epsilon \to 0+} \sum_{n=1}^{\infty} e^{-\epsilon n^2} \frac{\sin nx}{n} = \sum_{n=1}^{\infty} \frac{\sin nx}{n},$$

so that (4) is established. Equation (5) follows by change of variable, and the lemma is proved.

Lemma 10.b

1. $0 < x < \pi$, $0 < t$

$$\Rightarrow \quad x = \int_0^\pi [\theta(x-y,t) - \theta(x+y,t)]y\,dy + \pi\int_0^t \varphi(\pi-x, t-y)\,dy.$$

From (1) we have as before

$$\int_0^\pi [\theta(x-y,t) - \theta(x+y,t)]y\,dy = 2\sum_{n=1}^\infty (-1)^{n+1} e^{-n^2 t} \frac{\sin nx}{n}. \quad (6)$$

From (5) we have

$$\pi\int_0^t \varphi(\pi-x, y)\,dy = 2\sum_{n=1}^\infty (-1)^{n+1} \frac{1 - e^{-n^2 t}}{n} \sin nx. \quad (7)$$

Adding equations (6) and (7), we obtain

$$2\sum_{n=1}^\infty (-1)^{n+1} \frac{\sin nx}{n},$$

a Fourier series whose sum is known to be x for $0 < x < \pi$. This completes the proof.

Now to be assured of uniqueness we define the boundary functions $f(x)$, $g_1(t)$, and $g_2(t)$ of Problem I so as to form a continuous set of values over the sides and base of the rectangle R,

$$R = \{(x,t) \mid 0 \leq x \leq \pi,\ 0 \leq t \leq c\}.$$

Theorem 10 Let $f(x) \in C$ on $0 \leq x \leq \pi$, $g_1(t)$ and $g_2(t) \in C$ on $0 \leq t < \infty$, $g_1(0) = f(0)$, $g_2(0) = f(\pi)$. Then

1. $u(x,t) \in H$ on $0 < x < \pi$, $0 < t < \infty$,
2. $\lim_{t \to 0+} u(x,t) = f(x)$ uniformly on $0 \leq x \leq \pi$,
3. $\lim_{x \to 0+} u(x,t) = g_1(t)$ uniformly on $0 \leq t \leq c$, every $c > 0$,
4. $\lim_{x \to \pi-} u(x,t) = g_2(t)$ uniformly on $0 \leq t \leq c$

$$\Leftrightarrow \quad u(x,t) = \int_0^\pi [\theta(x-y,t) - \theta(x+y,t)]f(y)\,dy$$

$$+ \int_0^t \varphi(x, t-y)g_1(y)\,dy + \int_0^t \varphi(\pi-x, t-y)g_2(y)\,dy. \quad (8)$$

10. UNIQUENESS

Set

$$\Delta(x) = \frac{f(\pi) - f(0)}{\pi} x + f(0),$$

so that $f(x) - \Delta(x) \in C$ on $0 \leq x \leq \pi$ and vanishes at 0 and π. By the lemmas

$$\Delta(x) = \int_0^\pi [\theta(x - y, t) - \theta(x + y, t)] \Delta(y) \, dy$$

$$+ f(0) \int_0^t \varphi(x, t - y) \, dy + f(\pi) \int_0^t \varphi(\pi - x, t - y) \, dy.$$

Since $f(0) = g_1(0)$, $f(\pi) = g_2(0)$, we have

$$u(x, t) - \Delta(x) = \int_0^\pi [\theta(x - y, t) - \theta(x + y, t)][f(y) - \Delta(y)] \, dy$$

$$+ \int_0^t \varphi(x, t - y)[g_1(y) - g_1(0)] \, dy$$

$$+ \int_0^t \varphi(\pi - x, t - y)[g_2(y) - g_2(0)] \, dy. \tag{9}$$

We can now show that the function (8) satisfies conditions 1–4. Consider the first integral (9). By Corollary 6.2, it tends uniformly to $f(x) - \Delta(x)$ on $0 \leq x \leq \pi$ as $t \to 0$ and of course tends uniformly to 0 on the vertical sides of R by Corollary 6.1 and Theorem 6.2. The second integral tends to $g_1(t) - g_1(0) = g_1(t) - \Delta(0)$ uniformly on $0 \leq t \leq c$ as $x \to 0+$; it approaches 0 uniformly on the base and the other vertical side, $x = \pi$, of R, by Theorem 7.3. The behavior of the third integral is obtained from that of the second by an obvious change of variable.

Thus

$$\lim_{t \to 0+} [u(x, t) - \Delta(x)] = f(x) - \Delta(x) \quad \text{uniformly on} \quad 0 \leq x \leq \pi,$$

from which 2 follows by cancellation. Since it is trivial that

$$\lim_{x \to 0+} \Delta(x) = \Delta(0) \quad \text{uniformly on} \quad 0 \leq t \leq c,$$

conclusion 3 follows in an obvious way. Similarly for 4.

Since conditions 1–4 allow us to apply Theorem 2 (Chapter II), there is only one solution. Therefore, those conditions are necessary and sufficient for the representations (8).

Corollary 10

$$1. \quad u(x, t) \in H \quad \text{on} \quad 0 \leq x \leq \pi, \quad 0 \leq t < \infty$$

$$\Rightarrow \quad u(x, t) = \int_0^\pi [\theta(x - y, t) - \theta(x + y, t)] u(y, 0) \, dy$$

$$+ \int_0^t \varphi(x, t - y) u(0, y) \, dy$$

$$+ \int_0^t \varphi(\pi - x, t - y) u(\pi, y) \, dy.$$

Here $u(x, t)$ is continuous in the closed rectangle R so that the representation (8) is valid using the values of $u(x, t)$ on the appropriate boundaries.

Chapter VI

GREEN'S FUNCTION

1 GREEN'S FUNCTION FOR A RECTANGLE

A Green's function for a rectangle

$$R = \{(x, t) \mid 0 \leq x \leq \pi, \ 0 \leq t \leq c\}$$

with boundary Γ is one which may be used as the kernel of an integral transform which will represent an arbitrary temperature function inside R in terms of its values on Γ. We have already met such a representation in Corollary 10, Chapter V. Here we shall proceed axiomatically, later making contact with that earlier formula. Denote the base and vertical sides of R by γ.

Definition 1 The Green's function $G(x, t; \xi, \eta)$ for R is, for each fixed (ξ, η) inside R, a function of (x, t) such that

A. $G(x, t; \xi, \eta) \in H$ throughout R except at (ξ, η);
B. $G(x, t; \xi, \eta) = 0,$ (x, t) on γ;
C. $G(x, t; \xi, \eta) - k(x - \xi, t - \eta) \in H$ throughout R [if properly defined for $(x, t) = (\xi, \eta)$].

For such a function to exist it is clear that G must have the same type of singularity at (ξ, η) as does $k(x - \xi, t - \eta)$. There can be, at most, *one* Green's function, for, if there were two, their difference would belong to H throughout R and would vanish on γ (and thus throughout R by Theorem 2, Chapter II). We show now that G exists.

Theorem 1

$$G(x, t; \xi, \eta) = \theta(x - \xi, t - \eta) - \theta(x + \xi, t - \eta). \qquad (1)$$

Recall that

$$\theta(x - \xi, t - \eta) - \theta(x + \xi, t - \eta)$$

$$= \frac{2}{\pi} \sum_{n=1}^{\infty} e^{-n^2(t-\eta)} \sin n\xi \sin nx, \qquad t > \eta, \qquad (2)$$

$$= 0, \qquad t < \eta.$$

The θ-function (1) has its only singularities at $(x, t) = (\pm\xi + 2n\pi, \eta)$, $n = 0, \pm 1, \pm 2, \ldots$, by Theorem 3, Chapter V. It vanishes on γ by inspection of (2). Also

$$\theta(x - \xi, t - \eta) - \theta(x + \xi, t - \eta) - k(x - \xi, t - \eta)$$
$$= -\theta(x + \xi, t - \eta) + \sum_{|n| \geq 1} k(x - \xi + 2n\pi, t - \eta). \qquad (3)$$

The first function on the right has its only singularities at $(-\xi + 2n\pi, \eta)$, $n = 0, \pm 1, \pm 2, \ldots$, and hence none in R. Finally, the series (3) has its only singularities at $(\xi + 2n\pi, \eta)$, $n = \pm 1, \pm 2, \ldots$ and hence none in R. That is, the θ-function (1) satisfies A–C and hence must be the Green's function. The proper definition referred to in C is, of course, the value of the function on the right of (3) at (ξ, η).

From this explicit formula for G we may read off its properties as a function of (ξ, η) for fixed (x, t) inside R. Denote the top and vertical sides of R by γ^*.

Corollary 1 For fixed (x, y) inside R, $G(x, y; \xi, \eta)$ as a function of (ξ, η) satisfies the following conditions:

A. $G(x, t; \xi, \eta) \in H^*$ throughout R except at (x, t);
B. $G(x, t; \xi, \eta) = 0$, (ξ, η) on γ^*;
C. $G(x, t; \xi, \eta) - k(x - \xi, t - \eta) \in H^*$ throughout R [if properly defined at (x, t)].

2 AN INTEGRAL REPRESENTATION

As predicted in §1 we can now use the Green's function to obtain an integral representation of a temperature function in terms of its values on the boundary Γ of a rectangle R with boundary Γ:

$$R = \{(x, t) \mid 0 \leqslant x \leqslant \pi, \ 0 \leqslant t \leqslant c\}.$$

Theorem 2

1. $u(x, t) \in H$ throughout R,
2. (x, t) interior to R

$\Rightarrow \quad u(x, t) = \int_\Gamma G(x, y; \xi, \eta) u(\xi, \eta) \, d\xi$

$\qquad + [G(x, t; \xi, \eta) u_\xi(\xi, \eta) - G_\xi(x, t; \xi, \eta) u(\xi, \eta)] \, d\eta,$ (1)

where integration is counterclockwise over Γ.

It should be noted at once that the line integral (1) gives only zero contribution for integration along the top of R since G, considered as a function of (ξ, η), is identically zero there by B of Corollary 1. For fixed (x, t) set

$$v(\xi, \eta) = G(x, t; \xi, \eta) - k(x - \xi, t - \eta), \qquad (2)$$

so that $v(\xi, \eta) \in H^*$ in R by C, Corollary 1. If G is replaced by v in integral (1), the value of the integral is 0 by Theorem 4, Chapter I. Hence, integral (1) is equal to

$\int_\Gamma k(x - \xi, t - \eta) u(\xi, \eta) \, d\xi + [k(x - \xi, t - \eta) u_\xi(\xi, \eta)$

$\qquad - \tfrac{1}{2} h(x - \xi, t - \eta) u(\xi, \eta)] \, d\eta.$ (3)

For then by (2), G must be replaced by k and G_ξ by $-k' = h/2$. But integral (3) has the value $u(x, t)$ by Corollary 12, Chapter IV. This completes the proof.

Corollary 2a

$$u(x, t) = \int_\Gamma G(x, t; \xi, \eta) u(\xi, \eta) \, d\xi - G_\xi(x, t; \xi, \eta) u(\xi, \eta) \, d\eta. \quad (4)$$

This follows from (1) by noting that $G = 0$ for (ξ, η) on vertical sides of R and $d\eta = 0$ on horizontal sides. Equation (4) is important for the theory since the representation is in terms of $u(\xi, \eta)$ and not of $u_\xi(\xi, \eta)$.

Corollary 2b

$$u(x, t) = \int_0^\pi G(x, t; \xi, 0) u(\xi, 0) \, d\xi$$

$$+ \int_0^t G_\xi(x, t; 0, \eta) u(0, \eta) \, d\eta - \int_0^t G_\xi(x, t; \pi, \eta) u(\pi, \eta) \, d\eta. \quad (5)$$

This is an explicit statement of equation (4).

Corollary 2c

$$u(x, t) = \int_0^\pi [\theta(x - \xi, t) - \theta(x + \xi, t)] u(\xi, 0) \, d\xi$$

$$+ \int_0^t \varphi(x, t - \eta) u(0, \eta) \, d\eta + \int_0^t \varphi(\pi - x, t - \eta) u(\pi, \eta) \, d\eta. \quad (6)$$

This is a restatement of (5), using Theorem 1. Clearly

$$G_\xi = -\theta'(x - \xi, t - \eta) - \theta'(x + \xi, t - \eta)$$

$$= \tfrac{1}{2}\varphi(x - \xi, t - \eta) + \tfrac{1}{2}\varphi(x + \xi, t - \eta),$$

$$G_\xi(x, t; 0, \eta) = \tfrac{1}{2}\varphi(x, t - \eta) + \tfrac{1}{2}\varphi(x, t - \eta) = \varphi(x, t - \eta),$$

$$G_\xi(x, t; \pi, \eta) = \tfrac{1}{2}\varphi(x - \pi, t - \eta) + \tfrac{1}{2}\varphi(x + \pi, t - \eta)$$

$$= \varphi(x - \pi, t - \eta) = -\varphi(\pi - x, t - \eta).$$

Here we have used the periodicity of φ. In this corollary we have recaptured the conclusion of Corollary 10, Chapter V.

3 PROBLEM I AGAIN

Let us solve the first boundary-value problem for a rectangle R in terms of the Green's function, assuming that the given functions present a continuous set of values over the partial boundary γ.

3. PROBLEM I AGAIN

Theorem 3 Let $f(x) \in C$ on $0 \leq x \leq \pi$, $g_1(t)$ and $g_2(t) \in C$ on $0 \leq t < \infty$, $g_1(0) = f(0)$, and $g_2(0) = f(\pi)$. Then

1. $u(x, t) \in H$ on $0 < x < \pi$, $0 < t < \infty$,
2. $\lim_{t \to 0+} u(x, t) = f(x)$ uniformly on $0 \leq x \leq \pi$,
3. $\lim_{x \to 0+} u(x, t) = g_1(t)$ uniformly on $0 \leq t \leq c$, every c,
4. $\lim_{x \to \pi-} u(x, t) = g_2(t)$ uniformly on $0 \leq t \leq c$

$\Leftrightarrow u(x, t) = \int_0^\pi G(x, t; \xi, 0) f(\xi)\, d\xi$

$+ \int_0^t G_\xi(x, t; 0, \eta) g_1(\eta)\, d\eta - \int_0^t G_\xi(x, t; \pi, \eta) g_2(\eta)\, d\eta. \quad (1)$

As in the proof of Theorem 10, Chapter V, we see that there is, at most, one function $u(x, t)$ satisfying conditions 1–4. We need only show that the function (1) satisfies them. The temperature function

$$\Delta(x) = \frac{f(\pi) - f(0)}{\pi} x + f(0)$$

must have the representation 2(5):

$\Delta(x) = \int_0^\pi G(x, t; \xi, 0) \Delta(\xi)\, d\xi$

$+ \int_0^t G_\xi(x, t; 0, \eta) g_1(0)\, d\eta - \int_0^t G_\xi(x, t; \pi, \eta) g_2(0)\, d\eta.$

Hence,

$u(x, t) - \Delta(x) = \int_0^\pi G(x, t; \xi, 0)[f(\xi) - \Delta(\xi)]\, d\xi$

$+ \int_0^t G_\xi(x, t; 0, \eta)[g_1(\eta) - g_1(0)]\, d\eta$

$- \int_0^t G_\xi(x, t; \pi, \eta)[g_2(\eta) - g_2(0)]\, d\eta.$

As in Corollary 2c, these kernels can be expressed in terms of θ-functions, after which the proof proceeds as in that of Theorem 10, Chapter V.

It is noteworthy that in the presence of Theorem 2, we have avoided the tedious proofs of Lemmas 10a and 10b, Chapter V.

4 A PROPERTY OF $G(x, t; \xi, \eta)$

We now show that the Green's function $G(x, t; \xi, \eta)$, as a function of (x, t), represents the temperature of an infinite bar along the x-axis, initially at temperature 0, into which at the instant $t = \eta$ have been inserted sources of strength 1 at points $x = \xi + 2n\pi$ and sources of strength -1 (*sinks*) at points $x = -\xi + 2n\pi$, $n = 0, \pm 1, \pm 2, \ldots$. By periodicity, it will be sufficient to prove

Theorem 4

1. $0 < \epsilon < \xi, \qquad \epsilon < \pi - \xi$

$\Rightarrow$

A. $\quad \lim\limits_{t \to \eta+} \int_{\xi-\epsilon}^{\xi+\epsilon} G(x, t; \xi, \eta) \, dx = 1,$ (1)

B. $\quad \lim\limits_{t \to \eta+} \int_{-\xi-\epsilon}^{-\xi+\epsilon} G(x, t; \xi, \eta) \, dx = -1.$ (2)

By Theorem 1, equation (1) is equivalent to

$$\lim_{t \to 0+} \int_{\xi-\epsilon}^{\xi+\epsilon} [\theta(x - \xi, t) - \theta(x + \xi, t)] \, dx$$

or to

$$\lim_{t \to 0+} \int_0^\pi [\theta(\xi - x, t) - \theta(\xi + x, t)] f(x) \, dx, \qquad (3)$$

where $f(x)$ is defined as 1 on $(\xi - \epsilon, \xi + \epsilon)$ and as 0 elsewhere. By Corollary 6.2a, Chapter V, this limit is $f(\xi) = 1$, as desired. Replacing x by $-x$ in the integral (2), we obtain

$$\int_{\xi-\epsilon}^{\xi+\epsilon} [\theta(x + \xi, t) - \theta(x - \xi, t)] \, dx$$

$$= -\int_{\xi-\epsilon}^{\xi+\epsilon} [\theta(\xi - x, t) - \theta(\xi + x, t)] \, dx.$$

The limit of this integral is -1, by (3), as desired.

5 GREEN'S FUNCTION FOR AN ARBITRARY RECTANGLE

In §1 the rectangle R had its position with respect to the axes and its width somewhat specialized. It is useful to convert our formulas to the case of an arbitrary rectangle with sides parallel to the axes:

$$R = \{(x, t) \mid a \leqslant x \leqslant b, \ c \leqslant x \leqslant d\}, \qquad b - a = l. \tag{1}$$

If the θ-function is to have period $2l$ instead of 2π, it is

$$\theta(x, t) = \sum_{n=-\infty}^{\infty} k(x + 2nl, t). \tag{2}$$

Expanding this function in Fourier series, we have

$$\theta(x, t) = \frac{1}{2l} \sum_{n=-\infty}^{\infty} e^{ni\pi x/l - n^2\pi^2 t/l^2}. \tag{3}$$

Defining $G(x, t; \xi, \eta)$ for the rectangle (1) as in §1, we obtain

$$G(x, t; \xi, \eta) = \theta(x - \xi, t - \eta) - \theta(x - 2a + \xi, t - \eta)$$

$$= \frac{2}{l} \sum_{n=1}^{\infty} e^{-n^2\pi^2(t-\eta)/l^2} \sin n\pi \frac{(x-a)}{l} \sin n\pi \frac{(\xi-a)}{l}, \qquad t > \eta,$$

$$= 0, \qquad t < \eta. \tag{4}$$

With this formula for G and the rectangle (1) bounded by Γ, Corollary 2a becomes:

Theorem 5

1. $u(x, t) \in H$ throughout R
2. (x, t) interior to R

$$\Rightarrow u(x, t) = \int_\Gamma G(x, t; \xi, \eta) u(\xi, \eta) \, d\xi - G_\xi(x, t; \xi, \eta) u(\xi, \eta) \, d\eta,$$

where integration over Γ is in the counterclockwise sense.

6 SERIES OF TEMPERATURE FUNCTIONS

It is a familiar fact that the sum of a uniformly convergent series of harmonic functions is itself harmonic. It seems to be less familiar that the corresponding result for temperature functions is also true. We prove the result here.

Theorem 6

$$\begin{aligned} &1.\ u_n(x, t) \in H \quad \text{in a domain } D, \quad n = 1, 2, \ldots \\ &2.\ u(x, t) = \sum_{n=1}^{\infty} u_n(x, t) \quad \text{converges uniformly in } D \end{aligned} \tag{1}$$

$$\Rightarrow \quad u(x, t) \in H \quad \text{in } D.$$

Let (x_1, t_1) be a point of D. Choose constants a, b, c, d so that the rectangle R of §5 lies in D and contains (x_1, t_1) as an interior point. By Theorem 5 we have for (x, t) interior to R

$$u_n(x, t) = \int_\Gamma G(x, t; \xi, \eta) u_n(\xi, \eta)\, d\xi - G_\xi(x, t; \xi, \eta) u_n(\xi, \eta)\, d\eta. \tag{2}$$

Since for fixed (x, t) interior to R, $G(x, t; \xi, \eta)$ and $G_\xi(x, t; \xi, \eta) \in C$ for (ξ, η) on Γ, and since the series

$$u(\xi, \eta) = \sum_{n=1}^{\infty} u_n(\xi, \eta)$$

converges uniformly on Γ by 2, we may multiply this series by $G(x, t; \xi, \eta)$ and then by $G_\xi(x, t; \xi, \eta)$ and integrate term by term over Γ to obtain

$$\int_\Gamma G(x, t; \xi, \eta) u(\xi, \eta)\, d\xi - G_\xi(x, t; \xi, \eta) u(\xi, \eta)\, d\eta$$

$$= \sum_{n=1}^{\infty} \int_\Gamma G(x, t; \xi, \eta) u_n(\xi, \eta)\, d\xi - G_\xi(x, t; \xi, \eta) u_n(\xi, \eta)\, d\eta. \tag{3}$$

This series is equal to

$$\sum_{n=1}^{\infty} u_n(x, t) = u(x, t)$$

by (1) and (2). But by Corollary 2c the line integral (3) is the sum of three θ-transforms each of which satisfies the heat equation inside R, and hence at (x_1, t_1), by Corollary 6.1 and Theorem 7.1, Chapter V. Since (x_1, t_1) was an arbitrary point of D, $u(x, t) \in H$ throughout D, as stated.

7 THE REFLECTION PRINCIPLE

The Schwarz reflection principle is familiar for harmonic functions and for analytic functions of the complex variable. It is less well-known that the principle applies to temperature functions. It states, for example, that if $u \in H$ to the right of a vertical line in the x, t-plane and vanishes on that line, then it can be extended as a function of H through that line to the left. In fact this is done in such a way that the extended function has opposite signs at pairs of points which are reflections of each other in the given line. There is no restriction in taking the line as the t-axis.

Theorem 7

1. $u(x, t) \in H,$ $\qquad 0 < x < \pi, \quad 0 < t < c,$
2. $u(x, t) \in C,$ $\qquad 0 \leqslant x \leqslant \pi, \quad 0 \leqslant t \leqslant c,$
3. $u(0, t) = 0,$ $\qquad\qquad\qquad\quad 0 \leqslant t \leqslant c,$
4. $u(x, t) = -u(-x, t),$ $\quad -\pi \leqslant x \leqslant 0, \quad 0 \leqslant t \leqslant c,$

$\Rightarrow \qquad u(x, t) \in H, \quad -\pi \leqslant x \leqslant \pi, \; 0 < t < c.$

For the rectangle R [equation 5(1)], choose $a = -\pi$, $b = \pi$, $l = 2\pi$, so that $\theta(x, t)$, 5(2), has period 4π in x. Using the functions (2)–(4) of §5, form the function

$$U(x, t) = \int_\Gamma G(x, t; \xi, \eta) u(\xi, \eta) \, d\xi - G_\xi(x, t; \xi, \eta) u(\xi, \eta) \, d\eta,$$

or explicitly, by Corollary 2c,

$$U(x, t) = \int_{-\pi}^{\pi} [\theta(x - \xi, t) - \theta(x + 2\pi + \xi, t)] u(\xi, 0) \, d\xi$$

$$+ \int_0^t \varphi(x + \pi, t - \eta) u(-\pi, \eta) \, d\eta$$

$$+ \int_0^t \varphi(\pi - x, t - \eta) u(\pi, \eta) \, d\eta.$$

Of course this φ-function also has period 4π. Each of these integrals represents a function of H inside R by Chapter V. Hence we need only show that $U = u$ in the right half of R. Note first that $U(0, t) = 0$. The first integral vanishes since $u(\xi, 0)$ is odd and $\theta(-\xi, t) - \theta(2\pi + \xi, t)$ is even [θ is even and $2\pi + \xi \equiv \xi - 2\pi$, mod 4π]. Also, by 4,

$$\int_0^t \varphi(\pi, t - \eta) u(-\pi, \eta) d\eta = -\int_0^t \varphi(\pi, t - \eta) u(\pi, \eta) d\eta.$$

Since $u(\xi, \eta) \in C$ for (ξ, η) on Γ, $U(x, t)$ approaches $u(\xi, \eta)$ uniformly as (ξ, η) approaches Γ, by Theorem 10, Chapter V. Hence both $U(x, t)$ and $u(x, t)$ are continuous in the closed rectangle consisting of the right half of R and have the same boundary values on the sides and base thereof. Consequently, we may be assured by the uniqueness theorem of Chapter II, §3, that $U(x, t) = u(x, t)$ in the right half of R, as stated.

This reflection principle was used in the proof of the uniqueness theorem for the semi-infinite rod. The proof of that result, Theorem 6.2, Chapter II, is now complete.

8 ISOLATED SINGULARITIES

Such temperature functions as $k(x, t)$, $h(x, t)$, and their derivatives have isolated singularities. Only one of these, $k(x, t)$ is bounded on one side. We show here that it is the only such function, except of course for constant multiples thereof and for additive functions of H [Widder, 1930]. We prove first a series of preliminary results. Let R be a rectangle bounded by Γ:

$$R = \{(x, t) \mid a \leqslant x \leqslant b, \ c \leqslant x \leqslant d\}.$$

When further rectangles are needed, we use subscripts on all constants.

Lemma 8a

1. $w(x, t) \in H$ throughout R except at (α, β);
2. $R_0 \subset R$;
3. (α, β) is interior to R_0

$\Rightarrow$
$$\int_\Gamma w \, dx + w_x \, dt = \int_{\Gamma_0} w \, dx + w_x \, dt,$$

where integration over Γ and Γ_0 is in the same sense.

8. ISOLATED SINGULARITIES

Since $w \in H$ in the region between Γ and Γ_0, this result follows from Green's theorem (Chapter I, §4) with $u = w$, $v = 1$. From this result, the above line integral has the same value for all curves Γ_0. Set

$$\int_\Gamma w \, dx + w_x \, dt = K, \tag{1}$$

where integration is clockwise. We shall see that the function w, as defined in the following lemmas has many of the properties of the source solution itself.

Lemma 8b

1. $w(x, t) \in H$ throughout R except at (α, β);
2. $w(x, t) = 0$, $t < \beta$;
3. $a < \alpha - \delta < \alpha + \delta < b$

$\Rightarrow$
$$\lim_{r \to 0+} \int_{\alpha-\delta}^{\alpha+\delta} w(x, \beta + r) \, dx = K.$$

Choosing R_0 of Lemma 8a as

$$R_0 = \{(x, t) \mid \alpha - \delta \leq x \leq \alpha + \delta, \ c_0 \leq t \leq \beta + r\}, \qquad c < c_0 < \beta,$$

we obtain from (1) by integrating over Γ_0

$$\int_{\alpha-\delta}^{\alpha+\delta} w(x, \beta + r) \, dx + \int_\beta^{\beta+r} w_x(a - \delta, t) \, dt$$

$$- \int_\beta^{\beta+r} w_x(a + \delta, t) \, dt = K.$$

Since the second and third integrals tend to 0 with r, the result is immediate. An obvious consequence is that

$$\lim_{r \to 0+} \int_{\alpha+\delta}^{b} w(x, \beta + r) \, dx = 0,$$

$$\lim_{r \to 0+} \int_{a}^{\alpha-\delta} w(x, \beta + r) \, dx = 0.$$

We now impose the additional hypothesis that $w(x, t)$ is bounded below.

Lemma 8c

1. $w(x, t) \in H$ throughout R except at (α, β);
2. $w(x, t) = 0,$ $t < \beta$;
3. $w(x, t) \geq -M$ throughout R, some M;
4. $v(x, t) \in C$ throughout R;
5. $a < \alpha - \delta < \alpha + \delta < b$

$\Rightarrow \quad \lim_{r \to 0+} \int_{\alpha-\delta}^{\alpha+\delta} w(x, \beta + r)v(x, \beta + r)\, dx = Kv(\alpha, \beta).$

Notice that $k(x - \alpha, t - \beta)$ satisfies hypotheses 1–3 and that the conclusion holds if w is replaced by k.

Observe first that by 1 and 2, $w(x, \beta) = 0$, $x \neq \alpha$, so that

$$\lim_{r \to 0+} \int w(x, \beta + r)v(x, \beta + r)\, dx = 0 \qquad (2)$$

if the integration is over any interval not including $x = \alpha$. By Lemma 8b, it will be sufficient to prove that

$$\lim_{r \to 0+} \int_{\alpha-\delta}^{\alpha+\delta} w(x, \beta + r)[v(x, \beta + r) - v(\alpha, \beta)]\, dx = 0,$$

or, to make use of hypothesis 3, that

$$\lim_{r \to 0+} \int_{\alpha-\delta}^{\alpha+\delta} \{[w(x, \beta + r) + M][v(x, \beta + r) - v(\alpha, \beta)]$$

$$- M[v(x, \beta + r) - v(\alpha, \beta)]\}\, dx = 0. \qquad (3)$$

Given $\epsilon > 0$, we determine s so that when $|x - \alpha| < s$, $r < s$, then

$$|v(x, \beta + r) - v(\alpha, \beta)| < \epsilon. \qquad (4)$$

Now divide the integral (3) into three others, $I_1(r)$, $I_2(r)$, $I_3(r)$, corresponding respectively to the three intervals $(\alpha - s, \alpha + s)$, $(\alpha + s, \alpha + \delta)$, $(\alpha - \delta, \alpha - s)$. By (4) and hypothesis 3,

$$|I_1(r)| \leq \epsilon \int_{\alpha-s}^{\alpha+s} [w(x, \beta + r) + M]\, dx + 2Ms\epsilon. \qquad (5)$$

By (2),

$$\lim I_2(r) = \lim I_3(r) = 0. \qquad (6)$$

8. ISOLATED SINGULARITIES

By Lemma 8b and equations (5) and (6),

$$\varlimsup_{r \to 0+} |I_1(r) + I_2(r) + I_3(r)| \leq \epsilon(K + 4Ms). \tag{7}$$

Since ϵ is arbitrary, the limit superior (7), and hence the limit (3), is zero.

Theorem 8

1. $u(x, t) \in H$ throughout R except at (α, β),
2. $u(x, t) \geq -M$ throughout R, some M

$\Rightarrow$ $\quad u(x, t) = KG(x, t; \alpha, \beta) + f(x, t),$

where G is the Green's function for R and $f(x, t) \in H$ in R.

Define $f(x, t)$ such that $f \in H$ in R and $f = u$ on γ (sides and base of R.) This is possible by Theorem 2. Set $w = u - f$. Then w also satisfies 1 and 2 and is 0 on γ. Consequently, $w \equiv 0$ for $t < \beta$ by Theorem 2, Chapter II. Let (x_1, t_1) be an arbitrary point of R distinct from (α, β). Surround it by a rectangle R_1 not including (α, β) (see Figure 9). Surround (α, β) by a rectangle R_0, disjoint from R_1. Both R_0 and R_1 are inside R.

Now apply Green's theorem (Chapter I, §4) to the region between the boundaries Γ, Γ_0, Γ_1 with the functions u and v of that theorem here equal to w and $G(x_1, t_1; \xi, \eta)$, respectively. It is clear that $w(\xi, \eta) \in H$ and $G(x_1, t_1; \xi, \eta) \in H^*$ in that region, so that the line integral of the theorem extended over $\Gamma + \Gamma_0 + \Gamma_1$ is 0. But

$$\int_\Gamma w(\xi, \gamma) G(x_1, t_1; \xi, \eta) \, d\xi + (Gw_\xi - G_\xi w) \, d\eta = 0$$

since $w = 0$ on γ, $G = 0$ on γ^* (sides and top) by Corollary 1, and $d\eta = 0$ on top and bottom. Consequently

$$\int_{\Gamma_0} wG \, d\xi + (Gw_\xi - G_\xi w) \, d\eta = -\int_{\Gamma_1} wG \, d\xi + (Gw_\xi - G_\xi w) \, d\eta, \tag{8}$$

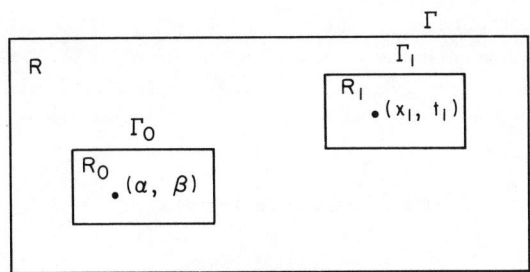

Figure 9

integration being clockwise, say, in both cases. We evaluate the first integral by Lemma 8c, the second by Corollary 12, Chapter IV. Using the notation for R_0 of Lemma 8b, the integral over Γ_0 is equal to

$$\int_{\alpha-\delta}^{\alpha+\delta} G(x_1, t_1; \xi, \beta + r) w(\xi, \beta + r) \, d\xi$$

$$+ \int_{\beta}^{\beta+r} \left\{ (Gw_\xi - G_\xi w)_{\xi=\alpha-\delta} - (Gw_\xi - G_\xi w)_{\xi=\alpha+\delta} \right\} d\eta.$$

When $r \to 0$ the second integral $\to 0$ and the first $\to KG(x_1, t_1; \alpha, \beta)$ by Lemma 8c.

For the integral over Γ_1 note first that it would be zero if G were replaced throughout by a function of H^*, by Green's theorem. Since, by Corollary 1, G differs from $k(x_1 - \xi, t_1 - \eta)$ by a function of H^* the value of the second integral (8) will be unchanged if G is replaced by $k(x_1 - \xi, t_1 - \eta)$ throughout. But then by Corollary 12, Chapter IV, its value is $w(x_1, t_1)$, so that equation (8) becomes

$$u(x_1, t_1) - f(x_1, t_1) = w(x_1, t_1) = KG(x_1, t_1; \alpha, \beta),$$

as we wished to prove. By definition of G we also have

$$u(x, t) = Kk(x - \alpha, t - \beta) + f(x, t), \tag{9}$$

where $f(x, t)$ is again a function of H in R.

This theorem enables us to classify isolated singularities of a temperature function. Suppose that $u(x, t) \in H$ in a neighborhood of (α, β) except perhaps at (α, β), so that hypothesis 1 of Theorem 8 holds for a suitable rectangle R. Then if $u(x, t)$ is bounded above and below, hypothesis 2, and hence equation (9), holds. But the right-hand side of (9) is unbounded unless $K = 0$. Then the singularity is removed by defining u at (α, β) as $f(\alpha, \beta)$. On the other hand, if $u(x, t)$ is bounded below or above only, then $u(x, t)$ has a source of strength K or a sink of strength $-K$ at (α, β) by equation (9).

As a consequence of the above considerations we see, for example, that all successive derivatives of $k(x, t)$ must be unbounded both above and below. For example, if $k'(x, t)$ were bounded below, then equation (9) would give

$$k'(x, t) = Kk(x, t) + f(x, t).$$

8. ISOLATED SINGULARITIES

Solving this linear differential equation, we would obtain for an arbitrary constant A,

$$k(x, t) = e^{Kx} \int e^{-Kx} f(x, t) \, dx + A e^{Kx}.$$

This is a contradiction since the right-hand side is continuous at the origin, whereas $k(x, t)$ is not. Of course the facts are evident from the explicit relation

$$\sqrt{4\pi t} \, k'(x, t) = -\frac{x}{2t} e^{-x^2/(4t)}.$$

Clearly $k' \to +\infty$ or $-\infty$ as $(x, t) \to (0, 0)$ from the left or from the right along the parabola $x^2 = 4t$. Since we have an explicit expression for $k^{(n)}$ in terms of the heat polynomials, we could check the conclusion in the general case.

Chapter VII

BOUNDED TEMPERATURE FUNCTIONS

1 THE INFINITE ROD

We are now in a position to study temperature functions which are Poisson transforms of bounded functions. We prove the elegant result that they are characterized by their boundedness alone, that is, $u(x, t) \in B \cdot H$ in the strip $-\infty < x < \infty$, $0 < t < c$ if and only if

$$u(x, t) = \int_{-\infty}^{\infty} k(x - y, t) f(y) \, dy, \qquad f(y) \in B(-\infty, \infty).$$

The proof will not involve some of the technical difficulties necessary to prove an analogous theorem for the positive functions of H which will be treated in the following chapter. Thus the essentials of the method will not be masked here by details.

We shall need as a tool the following result about weak compactness, the proof of which has been given, for example, by Banach [1932; 130].

Theorem A

1. $|f_n(x)| < M, \quad -\infty < x < \infty, \quad n = 0, 1, 2, \ldots, \quad$ some M

⇒ There exist integers $n_1, n_2, n_3, \ldots$ and $f(x) \in B(-\infty, \infty)$ such that

$$\lim_{k \to \infty} \int_{-\infty}^{\infty} g(x) f_{n_k}(x) \, dx = \int_{-\infty}^{\infty} g(x) f(x) \, dx$$

for every $g(x) \in L(-\infty, \infty)$.

We prove immediately the result mentioned above.

Theorem 1

1. $u(x, t) \in H \cdot B$, $\quad -\infty < x < \infty$, $\quad 0 < t < c$

⇔
$$u(x, t) = k * f = \int_{-\infty}^{\infty} k(x - y, t) f(y) \, dy, \tag{1}$$

some $f(y) \in B(-\infty, \infty)$.

We prove first the necessity of the conditions. Denote an upper bound for $|f(y)|$ by M. Then if integral (1) defines $u(x, t)$, we have

$$|u(x, t)| \leq M \int_{-\infty}^{\infty} k(x - y, t) \, dy = M, \quad 0 < t.$$

Thus $u(x, t)$ is well-defined and bounded in the half-plane $t > 0$. That it satisfies the heat equation there follows from Theorem 3, Chapter IV.

Conversely, we assume hypothesis 1 satisfied. Let (x, t) be an arbitrary point of the strip $0 < t < c$. For a fixed δ, $0 < \delta < c - t$, consider the integral

$$v(x, t) = \int_{-\infty}^{\infty} k(x - y, t) u(y, \delta) \, dy.$$

Since $u(y, \delta)$ is bounded, $v(x, t) \in B \cdot H$ for $t > 0$, as we proved above. Moreover, since $u(y, \delta) \in C$ on $-\infty < y < \infty$,

$$\lim_{t \to 0+} v(x, t) = u(x, \delta)$$

uniformly in every finite interval $|x| \leq R$, by Corollary 1, Chapter IV. Thus $v(x, t)$, like $u(x, t + \delta)$, $\in B \cdot H$ on $0 < t < c - \delta$ and $\in C$ on $0 \leq t < c - \delta$. Hence Theorem 6.1, Chapter II, is applicable to show

$$v(x, t) = u(x, t + \delta) = \int_{-\infty}^{\infty} k(x - y, t) u(y, \delta) \, dy, \quad 0 < t < c - \delta. \tag{2}$$

We now apply Theorem A to the bounded set of functions $u(y, \delta)$ where δ runs through a discrete set of numbers δ tending to zero. By that result, there exists a subset δ_n, $n = 1, 2, \ldots$, and a function $f(y) \in B(-\infty, \infty)$ such that

$$\lim_{n\to\infty} \int_{-\infty}^{\infty} k(x - y, t)u(y, \delta_n)\, dy = \int_{-\infty}^{\infty} k(x - y, t)f(y)\, dy. \tag{3}$$

For this to be a consequence of Theorem A, we must know that for fixed (x, t), $k(x - y, t) \in L(-\infty, \infty)$. But this is obvious from the equation

$$\int_{-\infty}^{\infty} k(x - y, t)\, dy = 1.$$

Since

$$\lim_{n\to\infty} u(x, t + \delta_n) = u(x, t),$$

equations (2) and (3) imply (1), as desired.

We point out that any function belonging to $B \cdot H$ in a strip $0 < t < c$ can be extended as a function of $B \cdot H$ to the half-plane $0 < t < \infty$, precisely by the transform (1).

Corollary 1

1. $u(x, t) \in H \cdot B$, $\quad -\infty < x < \infty$, $\quad 0 < t < c$,

$\Rightarrow \quad u(x, 0+) \quad$ exists almost everywhere $\quad (-\infty, \infty)$.

This follows from Corollary 5b, Chapter IV.

A simple illustration is provided by the function $e^{-t} \sin x$, which is obviously bounded for $t > 0$ and has the representation (1) with $f(y) = \sin y$. The functions $k(x, t)$ and $h(x, t)$ cannot have the representation for, although each is bounded on $-\infty < x < \infty$ for each $t > 0$, there is no uniform bound in the half-plane $t > 0$. This is clear because $k(0, 0+) = \infty$ and

$$\lim_{t\to 0+} h(\sqrt{t}, t) = \infty.$$

2 THE SEMI-INFINITE ROD

We may apply Theorem 1 to obtain a corresponding result for the semi-infinite rod.

Theorem 2

1. $u(x, t) \in H \cdot B$, $0 < x < \infty$, $0 < t < c$,
2. $u(x, t) \in C$, $0 \leq x < \infty$, $0 < t < c$,
3. $u(0, t) = 0$, $0 < t < c$,

$\Leftrightarrow$
$$u(x, t) = \int_0^\infty [k(x - y, t) - k(x + y, t)] f(y) \, dy,$$

some $f(y) \in B(0, \infty)$.   (1)

Assume first the representation (1). Define $f(y)$ for $y < 0$ to make it odd. Then (1) becomes

$$u(x, t) = \int_{-\infty}^\infty k(x - y, t) f(y) \, dy$$

with the extended function $f(y) \in B(-\infty, \infty)$. Thus, by Theorem 1, hypothesis 1 is here satisfied, while the validity of 2 and 3 may be seen by inspection. We could also infer the boundedness of $u(x,t)$ directly from equation (1) since, for $x > 0$, $y > 0$, $t > 0$,

$$k(x - y, t) - k(x + y, t) = 2k(x, t) e^{-y^2/(4t)} \sinh \frac{xy}{4t} > 0.$$

Conversely, if $u(x, t)$ satisfies conditions 1–3, we may extend $u(x, t)$ as a function of $H \cdot B$ into the negative half of the strip $0 < t < c$ by the definition

$$u(-x, t) = -u(x, t). \tag{2}$$

Here we have applied the reflection principle, Theorem 7, Chapter VI, valid in the presence of hypothesis 3. The extended function has the representation 1(1) by Theorem 1 for some $f(y) \in B$. By Corollary 5b, Chapter IV,

$$u(x, 0+) = f(x) \quad \text{almost all } x \, (-\infty, \infty),$$

and by (2) $f(x)$ is odd. Hence,

$$\int_{-\infty}^0 k(x - y, t) f(y) \, dy = -\int_0^\infty k(x + y, t) f(y) \, dy,$$

so that the integral 1(1) is equal to 2(1), as we wished to prove.

Here again $h(x, t)$ is excluded from the representation (1) even though $h(0, t) = 0$. All hypotheses of Theorem 2 are satisfied except $u \in B$.

3 SEMI-INFINITE ROD, CONTINUED

We consider next to the case, $u(x, t) \in H \cdot B$ in a half-strip, with $u(x, t)$ vanishing on the x-axis rather than on the t-axis.

Theorem 3

1. $u(x, t) \in H \cdot B$, $\quad 0 < x < \infty, \quad 0 < t < c$,
2. $u(x, t) \in C$, $\quad\quad\, 0 < x < \infty, \quad 0 \leqslant t < c$,
3. $u(x, 0) = 0$, $\quad\quad\;\, 0 < x < \infty$,

$$\Leftrightarrow \quad u(x, t) = \int_0^t h(x, t - y) f(y)\, dy, \quad \text{some } f(y) \in B(0, c). \tag{1}$$

Assume first the representation (1). By Theorem 7.1, Chapter IV,

$$|u(x, t)| \leqslant \int_0^t h(x, t - y) |f(y)|\, dy \leqslant \underset{0 < y < c}{\text{u.b.}} |f(y)|,$$

so that $u \in B$. By Theorem 8, Chapter IV, $u(x, t) \in H$. Since

$$|u(x, t)| \leqslant \underset{(0,c)}{\text{u.b.}} |f(y)| \frac{2}{\sqrt{\pi}} \int_{x/\sqrt{4t}}^{\infty} e^{-r^2}\, dr,$$

it is clear that for $\delta > 0$

$$\lim_{t \to 0+} u(x, t) = 0 \quad \text{uniformly}, \quad \delta \leqslant x < \infty,$$

so that conditions 2 and 3 are satisfied.

Conversely, we assume 1–3 and let (x, t) be an arbitrary point of the half-strip of hypothesis 1. For $\delta > 0$, set

$$v(x, t) = \int_0^t h(x, t - y) u(\delta, y)\, dy.$$

By Corollary 7.3, Chapter IV,

$$\lim_{x \to 0+} v(x, t) = u(\delta, t) \quad \text{uniformly}, \quad 0 \leqslant t < c,$$

$$\lim_{t \to 0+} v(x, t) = 0 \quad \text{uniformly}, \quad 0 \leqslant x < \infty.$$

The function $u(x + \delta, t)$ has the same property and hence must equal $v(x, t)$ by Theorem 6.2, Chapter II,

$$u(x + \delta, t) = \int_0^t h(x, t - y)u(\delta, y)\, dy.$$

Now apply Theorem A to the bounded set of functions $u(\delta, y)$, as was done in §1. It is valid, a fortiori, to the finite interval $(0, t)$. Consequently, a bounded function $f(y)$ and a sequence $\delta_1, \delta_2, \ldots,$ tending to zero, exist such that

$$u(x, t) = \lim_{k \to \infty} \int_0^t h(x, t - y)u(\delta_k, y)\, dy = \int_0^t h(x, t - y)f(y)\, dy.$$

Since the interval $(0, t)$ to which Theorem A is applied depends on t, it may at first appear that $f(y)$ also depends on t. That this is not the case follows from the uniqueness of the representation (1). If (1) held for two functions $f_1(y), f_2(y)$ we would have by Corollary 10.1a, Chapter IV,

$$0 = \lim_{x \to 0+} \int_0^t h(x, t - y)[f_1(y) - f_2(y)]\, dy = f_1(t) - f_2(t)$$

almost everywhere. This concludes the proof.

4 SEMI-INFINITE ROD, GENERAL CASE

In §§2 and 3, we treated the special cases in which $u(x, 0)$ or $u(0, t)$ vanished. We now proceed to the general case. For a preliminary result we add to the hypothesis $u \in H \cdot B$ the continuity of u on the two axes. This restriction will be removed later.

Theorem 4.1

$\quad$ 1. $u(x, t) \in H \cdot B, \quad 0 < x < \infty, \quad 0 < t < c,$
$\quad$ 2. $u(x, t) \in C, \quad\quad 0 \leqslant x < \infty, \quad 0 \leqslant t < c,$

$\Rightarrow \quad u(x, t) = \int_0^\infty [k(x - y, t) - k(x + y, t)]u(y, 0)\, dy$

$$+ \int_0^t h(x, t - y)u(0, y)\, dy. \tag{1}$$

Set $v(x, t)$ equal to the sum of the integrals (1). By 1 and 2, $u(y, 0)$ and $u(0, y)$ are bounded functions so that $v(x, t)$ satisfies hypothesis 1 by

Theorems 2 and 3. Observe next that (1) becomes the following familiar identity if $u(x, t) \equiv 1$:

$$1 = \frac{2}{\sqrt{\pi}} \int_0^x e^{-y^2} dy + \frac{2}{\sqrt{\pi}} \int_x^\infty e^{-y^2} dy.$$

Hence (1) is equivalent to

$$u(x, t) - u(0, 0) = \int_0^\infty [k(x - y, t) - k(x + y, t)]$$

$$\times [u(y, 0) - u(0, 0)] \, dy + \int_0^t h(x, t - y)[u(0, y) - u(0, 0)] \, dy. \quad (2)$$

Since $u(y, 0) - u(0, 0)$ vanishes at $y = 0$, its definition may be extended to negative y so as to be odd and to be continuous $-\infty < y < \infty$. Then the first integral (2) is equal to

$$\int_{-\infty}^\infty k(x - y, t)[u(y, 0) - u(0, 0)] \, dy,$$

and by Corollary 1, Chapter IV, $\to u(x, 0) - u(0, 0)$ uniformly in every finite interval as $t \to 0+$. It, of course, vanishes continuously on the t-axis. By Corollary 7.3, Chapter IV, the second integral (2) $\to u(0, t) - u(0, 0)$ uniformly on $0 \leq t < c$ as $x \to 0+$. It $\to 0$ uniformly on $0 \leq x < \infty$ as $t \to 0+$. Consequently, $v(x, t) - u(0, 0)$ and $u(x, t) - u(0, 0)$ both satisfy 1 and 2 and have equal values on the axes. Hence, by Theorem 6.2, Chapter II, they are equal and $u(x, t) = v(x, t)$, as we wished to prove.

Theorem 4.2

1. $u(x, t) \in H \cdot B, \quad 0 < x < \infty, \quad 0 < t < c,$

$\Leftrightarrow \quad u(x, t) = \int_0^\infty [k(x - y, t) - k(x + y, t)] f(y) \, dy$

$$+ \int_0^t h(x, t - y) g(y) \, dy \quad (3)$$

for some $f(y) \in B(0, \infty)$ and some $g(y) \in B(0, c)$.

Assuming the representation (3), each of the integrals defines a function

4. SEMI-INFINITE ROD, GENERAL CASE

of $H \cdot B$ by Theorems 2 and 3. Hence the necessity of condition 1 is obvious.

For the sufficiency, let (x, t) be an arbitrary point of the strip in question and let $0 < \delta < c - t$. Then $u(x + \delta, t + \delta)$ satisfies the conditions of Theorem 4.1 in the strip $0 < t < c - \delta$, so that

$$u(x + \delta, t + \delta) = \int_0^\infty [k(x - y, t) - k(x + y, t)] u(y + \delta, \delta) \, dy$$

$$+ \int_0^t h(x, t - y) u(\delta, y + \delta) \, dy. \quad (4)$$

Now apply Theorem A simultaneously to the bounded sets $u(y + \delta, \delta)$ and $u(\delta, y + \delta)$, δ tending to zero through some discrete set. We first establish the existence of an infinite subset for which the conclusion of Theorem A applies to the first integral (4). From this we pick a further subset for the second integral which will then apply to both. We thus conclude that there exists a sequence $\delta_1, \delta_2, \ldots$, tending to 0, a function $f(y) \in B\,(0, \infty)$ and a function $g(y) \in B\,(0, c)$ such that

$$\lim_{n \to \infty} u(x + \delta_n, t + \delta_n) = \lim_{n \to \infty} \left\{ \int_0^\infty [k(x - y, t) - k(x + y, t)] \right.$$

$$\left. \times u(y + \delta_n, \delta_n) \, dy + \int_0^t h(x, t - y) u(\delta_n, y + \delta_n) \, dy \right\}$$

$$u(x, t) = \int_0^\infty [k(x - y, t) - k(x + y, t)] f(y) \, dy$$

$$+ \int_0^t h(x, t - y) g(y) \, dy, \quad 0 < t < c - \delta. \quad (5)$$

The arbitrary point chosen is in the strip where (5) holds, so that the proof is complete.

Corollary 4.2

1. $u(x, t) \in H \cdot B, \quad 0 < x < \infty, \quad 0 < t < c,$

$\Rightarrow \; u(x, 0+)$ and $u(0+, t)$ exist almost everywhere on $(0, \infty)$ and $(0, c)$, respectively.

This is a consequence of Corollaries 5b and 10.1a, Chapter IV.

5 THE FINITE ROD

Here we show that any bounded temperature function on an open rectangle has an integral representation there. For closed rectangles this follows from Corollary 10, Chapter V, but here we admit the possibility of singularities on the boundary.

Theorem 5

1. $u(x, t) \in H \cdot B, \quad 0 < x < \pi, \quad 0 < t < c,$

$$\Leftrightarrow u(x, t) = \int_0^\pi [\theta(x - y, t) - \theta(x + y, t)] f(y) \, dy$$

$$+ \int_0^t \varphi(x, t - y) g_1(y) \, dy + \int_0^t \varphi(\pi - x, t - y) g_2(y) \, dy, \tag{1}$$

where

$$f(y) \in B(0, \pi), \quad g_1(y) \in B(0, c), \quad g_2(y) \in B(0, c).$$

Assuming equation (1), we see that $u(x, t) \in H$ by Corollary 6.1 and Theorem 7.1, Chapter V. If M is an upper bound for the absolute values of f, g_1, g_2 on their intervals of definition, then

$$|u(x, t)| \leq M \left\{ \int_0^\pi [\theta(x - y, t) - \theta(x + y, t)] \, dy \right.$$

$$\left. + \int_0^t \varphi(x, t - y) \, dy + \int_0^t \varphi(\pi - x, t - y) \, dy \right\}, \tag{2}$$

since the kernels are positive by Theorems 5.1 and 5.2, Chapter V. But the sum of the three integrals (2) is identically 1 by Corollary 10, Chapter V, applied to $u(x, t) \equiv 1$. Hence, $u \in H \cdot B$, as stated.

Conversely, let (x, t) be an arbitrary point in the given rectangle. Now choose δ so that $0 < \delta < \pi/2$ and $0 < \delta < c/2$, and also so small that $t < (c - 2\delta)/(1 - (2\delta/\pi))^2$. Then the function

$$u_\delta(x, t) = u\left(x(1 - 2\delta\pi^{-1}) + \delta, t(1 - 2\delta\pi^{-1})^2 + \delta\right) \in H$$

5. THE FINITE ROD

in the rectangle $0 \leq x \leq \pi$, $0 \leq t \leq (c - 2\delta)/(1 - 2\delta\pi^{-1})^2$. Since it belongs to H on the base and sides thereof, we may apply Corollary 10, Chapter V, to obtain

$$u_\delta(x, t) = \int_0^\pi [\theta(x - y, t) - \theta(x + y, t)] u_\delta(y, 0) \, dy$$

$$+ \int_0^t \varphi(x, t - y) u_\delta(0, y) \, dy$$

$$+ \int_0^t \varphi(\pi - x, t - y) u_\delta(\pi, y) \, dy. \tag{3}$$

Now apply Theorem A simultaneously to the bounded sets $u_\delta(y, 0)$, $u_\delta(0, y)$, and $u_\delta(\pi, y)$, with δ tending to zero through some discrete set. Thus, as in §4, we are assured of a subsequence $\delta_1, \delta_2, \ldots$ and functions $f(y) \in B(0, \pi)$, $g_1(y) \in B(0, c)$, $g_2(y) \in B(0, c)$ which are approached weakly by $u_{\delta_n}(y, 0)$, $u_{\delta_n}(0, y)$, $u_{\delta_n}(\pi, y)$, respectively, as $n \to \infty$. Since the three kernels of equation (1) surely belong to L on their respective intervals, Theorem A is applicable to give

$$\lim_{n \to \infty} u_{\delta_n}(x, t) = u(x, t)$$

$$= \lim_{n \to \infty} \left\{ \int_0^\pi [\theta(x - y, t) - \theta(x + y, t)] u_{\delta_n}(y, 0) \, dy \right.$$

$$\left. + \int_0^t \varphi(x, t - y) u_{\delta_n}(0, y) \, dy + \int_0^t \varphi(\pi - x, t - y) u_{\delta_n}(\pi, y) \, dy \right\}$$

$$= \int_0^\pi [\theta(x - y, t) - \theta(x + y, t)] f(y) \, dy$$

$$+ \int_0^t \varphi(x, t - y) g_1(y) \, dy + \int_0^t \varphi(\pi - x, t - y) g_2(y) \, dy.$$

The theorem is thus established.

Corollary 5

1. $u(x, t) \in H \cdot B$, $\quad 0 < x < \pi$, $\quad 0 < t < c$,

$\Rightarrow \quad u(x, 0 +)\quad\quad$ exists almost everywhere on $(0, \pi)$,

$\quad\quad u(0 +, t), u(\pi -, t) \quad$ exist almost everywhere on $(0, c)$.

Chapter VIII

POSITIVE TEMPERATURE FUNCTIONS

1 THE INFINITE ROD

In the previous chapter, we saw that bounded functions of H have simple integral representations. The basic functions $k(x, t)$ and $h(x, t)$ could not be included since they are unbounded in a neighborhood of the origin. However, $k(x, t)$ is bounded on one side since it is ≥ 0 everywhere. Also, $h(x, t) \geq 0$ in the half-plane $x > 0$. Here we shall develop integral expression for such functions, bounded on one side. For definiteness we assume them ≥ 0. Since a constant satisfies the heat equation, any results obtained can easily be applied to functions bounded on one side by any constant.

In place of Theorem A, Chapter VII, on weak compactness, we here need the selection principle of E. Helly [Widder, 1946; 29]. For the reader's convenience, we state the result here in the form we shall need.

Theorem A.

1. $\alpha_n(x) \in \uparrow$, $\quad -\infty < x < \infty, \quad n = 1, 2, \ldots,$
2. $|\alpha_n(x)| \leq M, \quad -\infty < x < \infty, \quad n = 1, 2, \ldots .$

⇒ There exist integers $n_1 < n_2 < \ldots$ and $f(x) \in \uparrow(-\infty, \infty)$ such that

$$\lim_{k \to \infty} \alpha_{n_k}(x) = f(x), \qquad -\infty < x < \infty, \qquad (1)$$

$$\lim_{k \to \infty} \int_{-\infty}^{\infty} \alpha_{n_k}(x) g(x)\, dx = \int_{-\infty}^{\infty} f(x) g(x)\, dx \qquad (2)$$

for every $g(x) \in L(-\infty, \infty)$.

For the proof of (1) see Widder [1946; 29]. Conclusion (2) is the familiar Lebesgue convergence theorem [Titchmarsh, 1939; 345].

2 UNIQUENESS, POSITIVE TEMPERATURES ON AN INFINITE ROD

The uniqueness result of Chapter II, Theorem 6.1, is not directly applicable here since hypothesis 3 is a two-sided condition. We wish now to establish a corresponding theorem in which that hypothesis is replaced by $u(x, t) \geq 0$. We need a preliminary result.

Theorem 2.1

1. $u(x, t) \in H, \geq 0, \qquad -\infty < x < \infty, \quad 0 < t < c,$
2. $0 < \delta < c$

⇒ $\int_{-\infty}^{\infty} k(x - y, t) u(y, \delta)\, dy \leq u(x, t + \delta), \qquad 0 < t < c - \delta. \quad (1)$

Later we shall be able to show that it is the equality rather than the inequality that holds in (1).

For an arbitrary number $A > 0$, consider the function

$$v(x, t) = u(x, t + \delta) - \int_{-A}^{A} k(x - y, t) u(y, \delta)\, dy \qquad (2)$$

on the rectangle

$$R = \{(x, t) \mid -B \leq x \leq B, \quad 0 \leq t \leq c - \delta\}, \qquad B > A.$$

As $t \to 0 +$, the integral (2) $\to u(x, \delta)$ uniformly on closed intervals inside $|x| < A$ and $\to 0$ uniformly on closed intervals outside $|x| \leq A$. It approaches no limit as $(x, t) \to (\pm A, 0)$ in the two-dimensional manner. But we can show that

$$\varlimsup_{(x, t) \to (A, 0)} \int_{-A}^{A} k(x - y, t) u(y, \delta) \, dy \leq u(A, \delta) \tag{3}$$

with a similar result for $-A$. For by continuity, given $\epsilon > 0$ there exists η such that $u(y, \delta) < u(A, \delta) + \epsilon$ on $(A - \eta, A)$. Hence the integral (3) is less than

$$[u(A, \delta) + \epsilon] \int_{A-\eta}^{A} k(x - y, t) \, dy. \tag{4}$$

This follows since the integrand (3) is nonnegative. Since the integral (4) is less than 1 and since $\in$ is arbitrary, inequality (3) follows immediately. These facts insure that

$$\varlimsup v(x, t) \geq 0$$

as $(x, t) \to$ any point on the base of R. On the sides, $x = \pm B$, the integral (2) can be made uniformly small by choice of B, since

$$\int_{-A}^{A} k(x - y, t) u(y, \delta) \, dy \leq \frac{1}{|x| - A} \int_{-A}^{A} u(y, \delta) \, dy, \qquad |x| > A.$$

Here we have used the trivial inequality $|x| e^{-x^2} < 1$ to show that $k(x, t) < 1/(\sqrt{\pi} |x|) < 1/|x|$. We conclude that for an arbitrary $\epsilon > 0$, $v(\pm B, t) \geq -\epsilon$ for large B. Thus we may apply Corollary 3.1, Chapter II, to $v(x, t) + \epsilon$ on R to see that $v(x, t) \geq -\epsilon$ inside R. Since ϵ was arbitrary, we conclude for any fixed (x, t) in the strip $0 < t < c - \delta$ that $v(x, t) \geq 0$ or that

$$\int_{-A}^{A} k(x - y, t) u(y, \delta) \, dy \leq u(x, t + \delta).$$

Since A was arbitrary, this also implies inequality (1), as we wished to prove.

We can now establish the main uniqueness result.

Theorem 2.2

1. $u(x, t) \in H, \geq 0,$ $\quad -\infty < x < \infty, \quad 0 < t < c,$
2. $u(x, t) \in C,$ $\quad -\infty < x < \infty, \quad 0 \leq t < c,$
3. $u(x, 0) = 0,$ $\quad -\infty < x < \infty$

$\Rightarrow \qquad u(x, t) \equiv 0, \quad -\infty < x < \infty, \quad 0 \leq t < c.$

2. UNIQUENESS, POSITIVE TEMPERATURES ON AN INFINITE ROD

Set

$$v(x, t) = \int_0^t u(x, y)\, dy.$$

By method H, (Chapter I, §6), $v(x, t) \in H$ for $0 < t < c$. Since $u(x, t) \geq 0$, $v(x, t) \in \uparrow$ as a function of t and is convex as a function of x. Also $v(x, t)$ satisfies hypothesis 3. If $v \equiv 0$ then $u \equiv 0$. Hence, there is no loss in generality if we add to our hypotheses that $u_t \geq 0$ and $u_{xx} \geq 0$.

Let $0 < \delta < c$ and $0 < t_0 < c - \delta$. Set

$$M = \sqrt{4\pi t_0} \int_{-\infty}^{\infty} k(y, t_0) u(y, \delta)\, dy.$$

By Theorem 2.1, $M \leq \sqrt{4\pi t_0}\, u(0, t_0 + \delta)$. For $x > 0$ and $|y| \leq 2x$,

$$\sqrt{4\pi t_0}\, k(y, t_0) = e^{-y^2/(4t_0)} \geq e^{-x^2/t_0},$$

and by convexity,

$$2xu(x, \delta) \leq \int_0^{2x} u(y, \delta)\, dy,$$

$$2xu(-x, \delta) \leq \int_{-2x}^{0} u(y, \delta)\, dy$$

(area under a convex arc $\geq$ area under any tangent). Combining these inequalities, we have

$$2xu(x, \delta) e^{-x^2/t_0} \leq \int_0^{2x} e^{-y^2/(4t_0)} u(y, \delta)\, dy \leq M.$$

Since $u_t \geq 0$, $u(\pm x, t) \leq u(\pm x, \delta)$ when $0 \leq t \leq \delta$, so that

$$u(\pm x, t) \leq \frac{Me^{x^2/t_0}}{2x} < Me^{x^2/t_0}, \qquad x > \tfrac{1}{2}.$$

Of course, $u(x, t)$ is bounded for $|x| \leq \tfrac{1}{2}$, so for suitable M_1

$$0 \leq u(x, t) \leq M_1 e^{x^2/t_0}, \qquad -\infty < x < \infty,\ 0 \leq t \leq \delta.$$

Hence we may apply Theorem 6.1, Chapter II, to conclude that $u \equiv 0$ for $0 \leq t \leq \delta$. Since δ was $< c$ but otherwise arbitrary, the proof is complete.

As stated earlier, we now see that it is the equality that holds in (1), for the function

$$u(x, t + \delta) - \int_{-\infty}^{\infty} k(x - y, t)u(y, \delta)\, dy$$

is ≥ 0 and satisfies all conditions of Theorem 2.2 for $0 < t < c - \delta$. It is consequently $\equiv 0$ there.

3 STIELTJES INTEGRAL REPRESENTATION, INFINITE ROD

We can now obtain a complete characterization of functions of H which are positive in a horizontal strip. They are those and only those which are Poisson–Stieltjes transforms of nondecreasing functions.

Theorem 3

1. $u(x, t) \in H, \geq 0, \qquad -\infty < x < \infty, \quad 0 < t < c,$

$$\Leftrightarrow \qquad u(x, t) = \int_{-\infty}^{\infty} k(x - y, t)\, d\alpha(y),$$

where $\alpha(y) \in \uparrow(-\infty, \infty)$.

This representation includes $k(x, t)$ itself, when $\alpha(y)$ is a step-function with a single jump at $y = 0$. The necessity of conditions 1 follows from Theorem 3, Chapter IV.

To prove the sufficiency, let (x_0, t_0) be an arbitrary point of the given strip and let $0 < \delta < c - t_0$. Form the function

$$\beta_\delta(x) = \int_{-\infty}^{x} k(y, t_0)u(y, \delta)\, dy \leq u(0, t_0 + \delta).$$

Here we have used 2(1), and by the same equation

$$u(x, t + \delta) = \int_{-\infty}^{\infty} k(x - y, t)u(y, \delta)\, dy = \int_{-\infty}^{\infty} \frac{k(x - y, t)}{k(y, t_0)}\, d\beta_\delta(y). \tag{1}$$

Obviously, $\beta_\delta(x) \in \uparrow$ and by the continuity of $u(0, t)$ at t_0

$$0 \leq \beta_\delta(x) \leq u(0, t_0) + 1, \qquad -\infty < x < \infty, \quad 0 < \delta < \delta_0,$$

for some δ_0. By Theorem A, there exist $\delta_1, \delta_2, \ldots$ approaching 0 and $\beta(x) \in \uparrow$ such that

$$\lim_{n \to \infty} \beta_{\delta_n}(x) = \beta(x).$$

Integrating (1) by parts, we have for $t < t_0$

$$u(x, t + \delta) = -\int_{-\infty}^{\infty} \beta_\delta(y) \frac{\partial}{\partial y} \frac{k(x - y, t)}{k(y, t_0)} dy, \qquad (2)$$

the integrated part vanishing since $\exp((y^2/t_0) - (y^2/t)) \to 0$, $y^2 \to \infty$. Similar considerations show that the derivative (2) $\in L(-\infty, \infty)$, so that 1(2) is applicable. We obtain

$$u(x, t) = -\int_{-\infty}^{\infty} \beta(y) \frac{\partial}{\partial y} \frac{k(x - y, t)}{k(y, t_0)} dy, \qquad 0 < t < t_0,$$

$$= \int_{-\infty}^{\infty} \frac{k(x - y, t)}{k(y, t_0)} d\beta(y). \qquad (3)$$

Finally, set

$$\alpha(x) = \int_0^x \frac{d\beta(y)}{k(y, t_0)},$$

so that equation (3) becomes

$$u(x, t) = \int_{-\infty}^{\infty} k(x - y, t) \, d\alpha(y), \qquad 0 < t < t_0.$$

Since $\alpha(x)$ is clearly nondecreasing and since t_0 was arbitrary, the theorem is proved. It may seem at first sight that $\alpha(y)$ depends on t_0. That it does not follows from the inversion formula of Chapter IV, Theorem 6.

By use of Theorem 3, it is possible to improve Theorem 2.2 by omitting hypothesis 2 and replacing 3 by $u(x, 0+) = 0$. That is, any nonnegative function $u(x, t) \in H$ which approaches 0 along every vertical line as $t \to 0+$ is identically zero. For the proof see Widder [1944; 94].

4 UNIQUENESS, SEMI-INFINITE ROD

In Chapter II, we were able to obtain the uniqueness result for the semi-infinite bar (Theorem 6.2) very simply from the previous one for the infinite bar, by use of the reflection principle. A similar attempt to extend the present Theorem 2.2 fails here since the reflected function is ≤ 0 for $x < 0$ and hypothesis 1 fails. Accordingly, we first obtain the analogue of Theorem 2.1.

Theorem 4.1

1. $u(x, t) \in H, \geq 0, \quad 0 < x < \infty, \quad 0 < t < c,$
2. $0 < \delta < c$

$\Rightarrow \int_0^\infty [k(x - y, t) - k(x + y, t)] u(y + \delta, \delta) \, dy$

$\leq u(x + \delta, t + \delta), \quad 0 < x < \infty, \quad 0 < t < c - \delta.$

For an arbitrary number $A > 0$, set

$$F(x, t) = \int_0^A [k(x - y, t) - k(x + y, t)] u(y + \delta, \delta) \, dy. \tag{1}$$

As in §2, $F(x, t) \to u(x + \delta, \delta)$ uniformly in closed intervals inside $0 < x < A$ and $\to 0$ uniformly in closed intervals on $A < x < \infty$. It approaches no limit as $(x, t) \to (A, 0)$ or $(0, 0)$. But as in §2

$$\varlimsup_{(x, t) \to (0, 0)} F(x, t) \leq u(\delta, \delta),$$

$$\varlimsup_{(x, t) \to (A, 0)} F(x, t) \leq u(A + \delta, \delta).$$

Now consider

$$v(x, t) = u(x + \delta, t + \delta) - F(x, t)$$

on the rectangle

$$R = \{(x, t) \mid 0 \leq x \leq B, \ 0 \leq t \leq c - \delta\}, \quad B > A.$$

From the above facts it is clear that

$$\varliminf v(x, t) \geq 0$$

as $(x, t) \to$ any point on the base of R. Since $k(x, t) \leq 1/|x|$

$$F(x, t) \leq \frac{2}{x - A} \int_0^A u(y + \delta, \delta) \, dy, \quad x > A, \tag{2}$$

and $F(B, t)$ is uniformly small on the right side of R when B is large. On the left side, $v(0, t) = u(\delta, t + \delta) \geq 0$. We may now conclude the proof as in §2 by applying Corollary 3.1, Chapter II, to R.

4. UNIQUENESS, SEMI-INFINITE ROD

Preliminary to the proof of the uniqueness theorem we establish the following result.

Lemma 4.2

$\qquad$ 1. $g(y) = k(1 - y, t_0) - k(1 + y, t_0), \qquad t_0 > 0,$

$\Rightarrow \qquad$ A. $g(y) \in \downarrow, \qquad\qquad y > R, \text{ some } R,$

$\qquad\qquad$ B. $\dfrac{1}{g(y)} = o(e^{ay^2}), \qquad y \to \infty, \quad a > \dfrac{1}{4t_0}.$

Differentiating $g(y)$, we have

$$2g'(y) = h'(1 - y, t_0) + h'(1 + y, t_0),$$

and this is negative when

$$(1 - y)e^{-(1-y)^2/(4t_0)} + (1 + y)e^{-(1+y)^2/(4t_0)} < 0$$

or when

$$e^{-y/t_0} < \frac{y - 1}{y + 1}.$$

The latter inequality is evident for large y, and A is proved. To prove B we have only to observe that

$$\lim_{y \to \infty} e^{y^2/(4t_0)} e^{-ay^2} = 0, \qquad a > \frac{1}{4t_0}.$$

Theorem 4.2

$\qquad$ 1. $u(x, t) \in H, \geq 0, \qquad 0 < x < \infty, \quad 0 < t < c,$
$\qquad$ 2. $u(x, t) \in C, \qquad\qquad 0 \leq x < \infty, \quad 0 \leq t < c,$
$\qquad$ 3. $u(x, 0) = 0, \qquad\qquad 0 \leq x < \infty,$
$\qquad$ 4. $u(0, t) = 0, \qquad\qquad 0 \leq t < c,$

$\Rightarrow \qquad u(x, t) \equiv 0, \qquad 0 \leq x < \infty, \quad 0 \leq t < c.$

As in the proof of Theorem 2.2, we may assume without loss of generality that $u_{xx} \geq 0$ and $u_t \geq 0$. Let $0 < \delta < c$ and $0 < t_0 < c - \delta$. By Theorem 4.1 and Lemma 4.2, we have for large x

$$g(3x) \int_x^{3x} u(y + \delta, \delta) \, dy \leq \int_x^{3x} g(y) u(y + \delta, \delta) \, dy$$

$$\leq u(1 + \delta, t_0 + \delta). \qquad (3)$$

Since $u_{xx} \geq 0$ and $u_t \geq 0$, we may proceed as in §2 to obtain

$$2xu(2x + \delta, t) \leq 2xu(2x + \delta, \delta)$$

$$\leq \int_x^{3x} u(y + \delta, \delta) \, dy, \qquad 0 < t < \delta. \qquad (4)$$

Combining (3) and (4) we have

$$u(2x + \delta, t) \leq \frac{u(1 + \delta, t_0 + \delta)}{2xg(3x)}.$$

Replacing $2x + \delta$ in this inequality by x and using B of Lemma 4.2 we obtain an exponential upper bound for $u(x, t)$ valid for large x. But by continuity we may then determine a constant M so large that

$$0 \leq u(x, t) \leq M e^{ax^2}, \qquad 0 \leq x < \infty.$$

Hence we are now in a position to apply Theorem 6.2, Chapter II, to conclude that $u(x, t) \equiv 0$, as stated.

5 REPRESENTATION, SEMI-INFINITE ROD

We first improve Theorem 4.1 as follows.

Theorem 5.1

    1. $u(x, t) \in H, \geq 0, \qquad 0 < x < \infty, \quad 0 < t < c,$
    2. $0 < \delta < c$

$\Rightarrow \qquad u(x, t) = \int_0^\infty [k(x - y, t) - k(x + y, t)] u(y + \delta, \delta) \, dy$

$$+ \int_0^t h(x, t - y) u(\delta, y + \delta) \, dy \leq u(x + \delta, t + \delta),$$

$$0 < x < \infty, \quad 0 < t < c - \delta.$$

Define $F(x, t)$ by 4(1) and set

$$G(x, t) = \int_0^t h(x, t - y) u(\delta, y + \delta) \, dy.$$

Consider the function

$$v(x, t) = u(x + \delta, t + \delta) - F(x, t) - G(x, t)$$

5. REPRESENTATION, SEMI-INFINITE ROD

on the rectangle
$$R = \{(x, t) \mid 0 \leqslant x \leqslant B, \; 0 \leqslant t \leqslant c - \delta\}, \quad B > A.$$
By Theorem 4.1,
$$\overline{\lim} \, F(x, t) \leqslant u(x + \delta, \delta)$$
as $(x, t) \to$ any point on the base of R. Since $G(x, 0+) = 0$ uniformly on closed intervals inside $0 < x < \infty$, we have
$$\underline{\lim} \, v(x, t) \geqslant 0 \tag{1}$$
as $(x, t) \to$ any point on the base of R except perhaps $(0, 0)$. By 4(2)
$$F(B, t) = O\left(\frac{1}{B}\right), \quad B \to \infty,$$
uniformly on $0 \leqslant t \leqslant c - \delta$. By Lemma 1, Chapter V
$$G(B, t) = O\left(\frac{1}{B^2}\right), \quad B \to \infty$$
uniformly on $0 \leqslant t \leqslant c - \delta$. These two relations show that for a given $\epsilon > 0$, B can be chosen so large that $v(x, t) \geqslant -\epsilon$ on the right-hand side of R. On the left-hand side $F(x, t) = 0$ and since
$$\lim_{x \to 0+} G(x, t) = u(\delta, t + \delta)$$
uniformly on closed intervals inside $0 < t < c - \delta$ we see that
$$\lim v(x, t) = 0$$
as $(x, t) \to$ any point on the left-hand side of R except perhaps $(0, 0)$. It remains to show that (1) holds when $(x, t) \to (0, 0)$, or equivalently
$$\overline{\lim_{(x, t) \to (0, 0)}} F(x, t) + G(x, t) \leqslant u(\delta, \delta).$$
To prove this write
$$F(x, t) = \int_0^A [k(x - y, t) - k(x + y, t)][u(y + \delta, \delta) - u(\delta, \delta)] \, dy$$
$$+ u(\delta, \delta) \int_0^A [k(x - y, t) - k(x + y, t)] \, dy, \tag{2}$$
$$G(x, t) = \int_0^t h(x, t - y)[u(\delta, y + \delta) - u(\delta, \delta)] \, dy$$
$$+ u(\delta, \delta) \int_0^t h(x, t - y) \, dy. \tag{3}$$

The first integral (2) → 0 by Corollary 1, Chapter IV, and the first integral (3) → 0 by Corollary 7.3, Chapter IV. The sum of the other two integrals is ≤

$$u(\delta, \delta) \int_0^\infty [k(x - y, t) - k(x + y, t)] \, dy$$

$$+ u(\delta, \delta) \int_0^t h(x, t - y) \, dy = u(\delta, \delta).$$

Here we have used Theorem 4.1, Chapter VII. Thus (1) holds on the entire side $x = 0$ and on the entire base $t = 0$; on $x = B$, $v(x, t) \geq -\epsilon$. Consequently, we may apply Corollary 3.1, Chapter II, to $v(x, t) + \epsilon$ on R to see that $v(x, t) \geq -\epsilon$ inside R. Since ϵ was arbitrary, we conclude for any fixed (x, t) in the strip $0 < t < c - \delta$ that $v(x, t) \geq 0$. Allowing A to become infinite, we obtain the desired conclusion.

Corollary 5.1

$$\int_0^\infty [k(x - y, t) - k(x + y, t)] u(y + \delta, \delta) \, dy$$

$$+ \int_0^t h(x, t - y) u(\delta, y + \delta) \, dy = u(x + \delta, t + \delta).$$

The right-hand side less the left-hand side is ≥ 0 by Theorem 5.1, satisfies all conditions of Theorem 4.2, and is consequently $\equiv 0$.

For the following representation theorem we need a preliminary result about Stieltjes integrals.

Lemma 5.2

1. $f(x), g(x) \in C, \quad > 0, \qquad\qquad 0 < x \leq 0,$
2. $f(x) \sim fx, \quad g(x) \sim gx, \quad b \neq 0, \quad x \to 0+,$
3. $\beta(x) \in \uparrow, \qquad\qquad 0 \leq x \leq 1,$
4. $\alpha(x) = -\int_x^1 \frac{d\beta(y)}{g(y)}$

⇒ A. $\int_0^1 \frac{f(x)}{g(x)} d\beta(x) = \frac{f}{g} [\beta(0+) - \beta(0)] + \int_{0+}^1 f(x) \, d\alpha(x),$

B. $\int_{0+}^1 x \, d\alpha(x) < \infty.$

We use familiar results [Widder, 1946; 12] to obtain for $0 < \epsilon < 1$

$$\int_0^1 \frac{f(x)}{g(x)} d\beta(x) = \int_0^\epsilon \frac{f(x)}{g(x)} d\beta(x) + \int_\epsilon^1 f(x) d\alpha(x). \tag{4}$$

Since $f(x)/g(x) \to f/g$ as $x \to 0+$ and since $\beta(x) \in \uparrow$, we may apply the law of the mean [Widder, 1946; 16] to obtain, for $0 < \xi < \epsilon$,

$$\lim_{\epsilon \to 0+} \int_0^\epsilon \frac{f(x)}{g(x)} d\beta(y) = \lim_{\epsilon \to 0+} \frac{f(\xi)}{g(\xi)} \int_0^\epsilon d\beta(x) = \frac{f}{g}[\beta(0+) - \beta(0)].$$

Allowing ϵ to approach 0 in (4), we obtain conclusion A. Conclusion B follows immediately from the relations

$$\int_\epsilon^1 x\, d\alpha(x) = \int_\epsilon^1 \frac{x}{g(x)} d\beta(x) \leqslant \int_{0+}^1 \frac{x}{g(x)} d\beta(x) < \infty.$$

Note that the integral B may well be improper. For example, if $\beta(x) = g(x) = x$, then $\alpha(0+) = -\infty$.

Theorem 5.2

1. $u(x, t) \in H, \geqslant 0, \quad 0 < x < \infty, \quad 0 < t < c,$

$\Leftrightarrow \quad u(x, t) = \int_{0+}^\infty [k(x - y, t) - k(x + y, t)]\, d\alpha(y)$

$$+ \int_0^t h(x, t - y)\, d\beta(y), \quad 0 < x < \infty, \quad 0 < t < c, \tag{5}$$

where

$$\alpha(y) \in \uparrow, \quad 0 < y < \infty, \quad \beta(y) \in \uparrow, \quad 0 \leqslant y < c.$$

Note that $\beta(0)$ is finite but that $\alpha(0+)$ may be $-\infty$. Of course the improper integral converges. For the necessity of conditions 1, we assume equation (5) and appeal to Theorems 3 and 8, Chapter IV.

To prove the converse, let (x, t) be an arbitrary point in the given half-strip. Choose a positive $\delta < c - t$ and choose a so that $t < a < c - \delta$. Set

$$\alpha_\delta(x) = \int_0^x g(y) u(y + \delta, \delta)\, dy,$$

$$\beta_\delta(x) = \int_0^x h(1, a - y) u(\delta, y + \delta)\, dy,$$

where
$$g(y) = k(1 - y, a) - k(1 + y, a). \tag{6}$$

Then by Corollary 5.1

$$u(x + \delta, t + \delta) = \int_0^\infty \frac{[k(x - y, t) - k(x + y, t)]}{g(y)} \, d\alpha_\delta(y)$$

$$+ \int_0^t \frac{h(x, t - y)}{h(1, a - y)} \, d\beta_\delta(y), \qquad 0 < t < a. \tag{7}$$

Moreover, by the same corollary,

$$\alpha_\delta(x) + \beta_\delta(x) \leq \int_0^\infty [k(1 - y, a) - k(1 + y, a)] u(y + \delta, \delta) \, dy$$

$$+ \int_0^a h(1, a - y) u(\delta, y + \delta) \, dy = u(1 + \delta, a + \delta).$$

Hence for δ in a neighborhood of $\delta = 0$ the sets of functions $\alpha_\delta(x)$ and $\beta_\delta(x)$ both have upper bounds independent of x and δ, so that we may apply Theorem A to be assured of the existence of a sequence $\delta_1, \delta_2, \ldots$ tending to 0 and two functions $\alpha^*(x) \in \uparrow$, $0 \leq x < \infty$, and $\beta^*(x) \in \uparrow$, $0 \leq x \leq a$, such that

$$\lim_{n \to \infty} \alpha_{\delta_n}(x) = \alpha^*(x), \qquad \lim_{n \to \infty} \beta_{\delta_n}(x) = \beta^*(x).$$

Integrating by parts in (7) as was done in §3 in order to apply Lebesgue's convergence theorem 1(2), we obtain

$$u(x, t) = \int_0^\infty \frac{k(x - y, t) - k(x + y, t)}{g(y)} \, d\alpha^*(y)$$

$$+ \int_0^t \frac{h(x, t - y)}{h(1, a - y)} \, d\beta^*(y), \qquad 0 < t < a. \tag{8}$$

To the first integral we now apply Lemma 5.1 with $g(y)$ defined by (6), $\beta(x)$ replaced by $\alpha^*(x)$, and for the fixed (x, t) chosen above

$$f(y) = k(x - y, t) - k(x + y, t).$$

Then
$$f = \lim_{y \to 0} \frac{k(x-y, t) - k(x+y, t)}{y} = -2k'(x, t) = h(x, t),$$

$$g = h(1, a) \neq 0,$$

$$\alpha(x) = -\int_x^1 \frac{d\alpha^*(y)}{g(y)},$$

and the first conclusion of the lemma yields

$$\int_0^\infty \frac{k(x-y, t) - k(x+y, t)}{g(y)} d\alpha^*(y)$$

$$= ph(x, t) + \int_{0+}^\infty [k(x-y, t) - k(x+y, t)] d\alpha(y),$$

$$p = \frac{[\alpha^*(0+) - \alpha^*(0)]}{h(1, a)}.$$

If we set

$$\beta(x) = \int_0^x \frac{d\beta^*(y)}{h(1, a-y)}, \tag{9}$$

equation (8) becomes

$$u(x, t) = ph(x, t) + \int_{0+}^\infty [k(x-y, t) - k(x+y, t)] d\alpha(y)$$

$$+ \int_0^t h(x, t-y) d\beta(y), \qquad 0 < t < a. \tag{10}$$

The term $ph(x, t)$ can be included in the second integral by altering the definition (9) of $\beta(x)$ by the addition of $pU(x)$, where $U(x)$ is the step-function with a single unit jump at $x = 0$. One sees by inspection that $\alpha(x)$ and $\beta(x)$ are nondecreasing, as stated. This concludes the proof. It is the vanishing of the kernel $k(x-y, t) - k(x+y, t)$ at $y = 0$ which permits $\alpha(0+)$ to be $-\infty$ and requires the nonintegral term in (10).

Corollary 5.2a

$$\int_{0+}^1 y \, d\alpha(y) < \infty.$$

This follows at once from conclusion B of Lemma 5.1. An example for which $\alpha(0+) = -\infty$ is provided by $\alpha(y) = -y^{-1/2}$. Then

$$u(x, t) = \frac{1}{2} \int_{0+}^{\infty} \frac{[k(x-y, t) - k(x+y, t)]}{y^{3/2}} \, dy$$

$$\int_{0+}^{1} y \, d\alpha(y) = \frac{1}{2} \int_{0+}^{1} \frac{dy}{\sqrt{y}} < \infty.$$

(11)

Special cases of the theorem occur when $u(x, t)$ vanishes on one of the axes. These are of sufficient interest to merit statement.

Corollary 5.2b

1. $u(x, t) \in H, \geq 0,$ $\quad 0 < x < \infty, \ 0 < t < c;$
2. $u(x, t) \in C,$ $\quad 0 \leq x < \infty, \ 0 < t < c;$
3. $u(0, t) = 0,$ $\quad 0 < t < c,$

$\Leftrightarrow u(x, t) = ph(x, t) + \int_{0+}^{\infty} [k(x-y, t) - k(x+y, t)] \, d\alpha(y),$

where $p \geq 0$ and $\alpha(y) \in \uparrow, 0 < y < \infty.$

Corollary 5.2c

1. $u(x, t) \in H, \geq 0,$ $\quad 0 < x < \infty, \ 0 < t < c;$
2. $u(x, t) \in C,$ $\quad 0 < x < \infty, \ 0 \leq t < c;$
3. $u(x, 0) = 0,$ $\quad 0 < x < \infty,$

$\Leftrightarrow u(x, t) = \int_{0}^{t} h(x, t-y) \, d\beta(y), \quad \beta(y) \in \uparrow, \ 0 \leq y < c.$

These results may be obtained from (10) by applying an inversion formula to show that one or another of the integrals disappears. We omit details.

The function (11) provides an illustration of Corollary 5.2b. To illustrate Corollary 5.2c we may use the basic functions $k(x, t)$ and $h(x, t)$ themselves. They both satisfy hypotheses 1–3, and

$$h(x, t) = \int_{0}^{t} h(x, t-y) \, dU(y)$$

where $U(y)$ is the unit step-function defined above. Also

$$k(x, t) = \int_0^t h(x, t - y) k(0, y)\, dy$$

$$= \int_0^t h(x, t - y) \frac{dy}{\sqrt{4\pi y}}, \quad x, t > 0. \tag{12}$$

To prove this we may appeal to Corollary 5.1 and allow δ to approach 0 under the integral signs. The step is easily justified by Lebesgue's convergence theorem. We give an alternative proof, using the product theorem for Laplace transforms, [Widder, 1971; 111]. By Theorems 7.2 and 8.1, Chapter III, the Laplace transforms of $k(x, t)$ and $h(x, t)$, considered as functions of t, are $e^{-x\sqrt{s}}/\sqrt{4s}$ and $e^{-x\sqrt{s}}$, respectively. Hence the Laplace transform of the convolution (12) is the product $e^{-x\sqrt{s}}$ by $1/\sqrt{4s}$. The determining function for the generating function $e^{-x/\sqrt{s}}/\sqrt{4s}$ is $k(x, t)$, as we just noted. This establishes equation (12).

6 THE FINITE ROD

For the representation theorem contemplated we need a preliminary result analogous to Lemma 5.1.

Lemma 6

1. $f(x), g(x) \in C, > 0,$ $0 < x < \pi,$
2. $f(x) \sim ax,\ g(x) \sim bx,\ b \neq 0,$ $x \to 0+,$
   $f(x) \sim c(\pi - x),\ g(x) \sim d(\pi - x),\ d \neq 0,$ $x \to \pi-,$
3. $\beta(x) \in \uparrow$ $0 \leq x \leq \pi,$
4. $\alpha(x) = \int_{\pi/2}^x \frac{d\beta(y)}{g(y)},$ $0 < x < \pi,$

$\Rightarrow$ A. $\displaystyle\int_0^\pi \frac{f(x)}{g(x)} d\beta(x) = \frac{a}{b}[\beta(0+) - \beta(0)]$
$$+ \frac{c}{d}[\beta(\pi) - \beta(\pi-)] + \int_{0+}^{\pi-} f(x)\, d\alpha(x),$$

B. $\displaystyle\int_{0+}^{\pi-} x(\pi - x)\, d\alpha(x) < \infty.$

By 2 the quotient $f(x)/g(x)$ may be defined at 0 and π to become continuous on $0 \leq x \leq \pi$, so that the first integral A exists. Hence,

$$\int_0^\pi \frac{f}{g}\,d\alpha = \int_0^\epsilon \frac{f}{g}\,d\alpha + \int_\epsilon^{\pi-\epsilon} \frac{f}{g}\,d\alpha + \int_{\pi-\epsilon}^\pi \frac{f}{g}\,d\alpha, \qquad 0 < \epsilon < \pi/2.$$

As $\epsilon \to 0$, the first and third integrals have limits equal to the first and second terms on the right of A. See Widder [1946; 9]. The middle integral must therefore approach a limit, the convergent improper integral in A.

To prove B we have

$$\int_\epsilon^{\pi-\epsilon} x(\pi - x)\,d\alpha(x) = \int_\epsilon^{\pi-\epsilon} \frac{x(\pi - x)}{g(x)}\,d\beta(x)$$

$$\leq \int_0^\pi \frac{x(\pi - x)}{g(x)}\,d\beta(x) < \infty.$$

The final integral exists since $x(\pi - x)/g(x) \in C$ on $0 \leq x \leq \pi$ if suitably defined at 0 and π. Conclusion B follows when $\epsilon \to 0$.

Theorem 6

1. $u(x, t) \in H, \geq 0, \qquad 0 < x < \pi, \quad 0 < t < c,$

$\Leftrightarrow \quad u(x, t) = \int_{0+}^{\pi-} [\theta(x - y, t) - \theta(x + y, t)]\,d\alpha(y)$

$$+ \int_0^t \varphi(x, t - y)\,d\beta(y) + \int_0^t \varphi(\pi - x, t - y)\,d\gamma(y), \quad (1)$$

where

$$\alpha(y) \in \uparrow, \qquad 0 < y < \pi, \qquad \beta(y), \gamma(y) \in \uparrow, \qquad 0 \leq y < c,$$

and

$$\int_{0+}^{\pi-} y(\pi - y)\,d\alpha(y) < \infty. \qquad (2)$$

Note that $\alpha(0+) = -\infty$ and $\beta(\pi-) = +\infty$ are possibilities but that $\beta(0)$ and $\gamma(0)$ are finite. If the open rectangle of hypothesis 1 were replaced by the corresponding closed rectangle, this theorem would be a trivial consequence of Corollary 10, Chapter V, with $\alpha'(y) = u(y, 0)$, $\beta'(y) = u(0, y)$, $\gamma'(y) = u(\pi, y)$. But under the present hypotheses we must again appeal to Helly's selection principle. The necessity of condition 1 is immediate from Corollary 6.1 and Theorem 7.1, Chapter V.

For the sufficiency, choose (x_1, t_1) in the given rectangle and set

$$u_\delta(x, t) = u\left(x\left[1 - \frac{2\delta}{\pi}\right] + \delta, \ t\left[1 - \frac{2\delta}{\pi}\right]^2 + \delta\right), \qquad 0 < \delta < \frac{\pi}{2}.$$

This function satisfies 1 for $0 \leq x \leq \pi, 0 \leq t \leq (c - 2\delta)/(1 - (2\delta/\pi))^2$, so that we may apply the result mentioned above:

$$u_\delta(x, t) = \int_0^\pi G(x, y, t) u_\delta(y, 0) \, dy$$

$$+ \int_0^t \varphi(x, t - y) u_\delta(0, y) \, dy + \int_0^t \varphi(\pi - x, t - y) u_\delta(\pi, y) \, dy, \quad (3)$$

$$G(x, y, t) = \theta(x - y, t) - \theta(x + y, t).$$

For δ sufficiently small, (x_1, t_1) is inside the rectangle of validity for (3). Now choose a fixed point (a, b), such that $0 < a < \pi$ and $t_1 < b < (c - 2\delta)/(1 - (2\delta/\pi))^2$. Set

$$\alpha_\delta(x) = \int_0^x G(a, y, b) u_\delta(y, 0) \, dy, \qquad 0 \leq x \leq \pi,$$

$$\beta_\delta(x) = \int_0^x \varphi(a, b - y) u_\delta(0, y) \, dy, \qquad 0 \leq x \leq b,$$

$$\gamma_\delta(x) = \int_0^x \varphi(\pi - a, b - y) u_\delta(\pi, y) \, dy, \qquad 0 \leq x \leq b.$$

By equation (3),

$$0 \leq \alpha_\delta(x) \leq u_\delta(a, b), \qquad 0 \leq x \leq \pi,$$

$$0 \leq \beta_\delta(x), \gamma_\delta(x) \leq u_\delta(a, b), \qquad 0 \leq x \leq b.$$

Since $u(x, t) \in C$ at (a, b), there exists δ_0 such that

$$0 \leq \alpha_\delta(x) \leq u(a, b) + 1, \qquad 0 \leq x \leq \pi, \ 0 < \delta < \delta_0,$$

$$0 \leq \beta_\delta(x), \gamma_\delta(x) \leq u(a, b) + 1, \qquad 0 \leq x \leq b, \ 0 < \delta < \delta_0.$$

Hence we may apply Theorem A to be assured of the existence of a sequence $\delta_1, \delta_2, \ldots$ tending to 0 and $\alpha^*(x) \in \uparrow$ on $0 \leq x \leq \pi$, $\beta^*(x)$ and $\gamma^*(x) \in \uparrow$ on $0 \leq x \leq b$ such that

$$\alpha_{\delta_n}(x) \to \alpha^*(x), \beta_{\delta_n}(x) \to \beta^*(x), \gamma_{\delta_n}(x) \to \gamma^*(x), \qquad n \to \infty.$$

Then, applying 1(2) after integrating by parts, to (3), we have

$$u_\delta(x, t) = \int_0^\pi \frac{G(x, y, t)}{G(a, y, b)} d\alpha_\delta(y) + \int_0^t \frac{\varphi(x, t-y)}{\varphi(a, b-y)} d\beta_\delta(y)$$

$$+ \int_0^t \frac{\varphi(\pi-x, t-y)}{\varphi(\pi-a, b-y)} d\gamma_\delta(y),$$

$$u(x, t) = \int_0^\pi \frac{G(x, y, t)}{G(a, y, b)} d\alpha^*(y) + \int_0^t \frac{\varphi(x, t-y)}{\varphi(a, b-y)} d\beta^*(y)$$

$$+ \int_0^t \frac{\varphi(\pi-x, t-y)}{\varphi(\pi-a, b-y)} d\gamma^*(y), \qquad (4)$$

for $0 < t < b$. Now apply Lemma 6, with primes on constants, to the first integral (4), here replacing $\beta(x)$ of the lemma by $\alpha^*(x)$ and setting

$$f(y) = G(x, y, t), \qquad a' = \lim_{y \to 0+} \frac{G(x, y, t)}{y} = \varphi(x, t), \qquad (5)$$

$$g(y) = G(a, y, b), \qquad b' = \varphi(a, b).$$

Here we have used L'Hospital's rule and $\theta'(x, t) = -\varphi(x, t)/2$. Since $\theta(-x, t) = \theta(x, t)$ and $\theta(x + 2\pi, t) = \theta(x)$, we obtain

$$G(x, y, t) = G(\pi - x, \pi - y, t),$$

so that we have from (5) that $c' = \varphi(\pi - x, t)$, $d' = \varphi(\pi - a, b)$. Thus by conclusion A of Lemma 6, equation (4) becomes

$$u(x, t) = \int_{0+}^{\pi-} G(x, y, t) \, d\alpha(y) + \int_0^t \varphi(x, t-y) \, d\beta(y)$$

$$+ \int_0^t \varphi(\pi - x, t - y) \, d\gamma(y) + p\varphi(x, t) + q\varphi(\pi - x, t), \qquad (6)$$

$$0 < x < \pi, \quad 0 < t < b,$$

where

$$p = \frac{\alpha^*(0+) - \alpha^*(0)}{\varphi(a, b)}, \qquad q = \frac{\alpha^*(\pi) - \alpha^*(\pi-)}{\varphi(\pi - a, b)},$$

$$\beta(x) = \int_0^x \frac{d\beta^*(y)}{\varphi(a, b-y)}, \qquad \gamma(x) = \int_0^x \frac{d\gamma^*(y)}{\varphi(\pi - a, b - y)}.$$

Clearly $p \geqslant 0$, $q \geqslant 0$, and $\alpha(x), \beta(x), \gamma(x) \in \uparrow$. By altering the definition of $\alpha(x)$ by the addition of the nondecreasing step-function $pU(x) + qU(x - \pi)$, the two nonintegral terms in (6) disappear. Since the chosen point (x_1, t_1) lies in the rectangle where (6) holds, equation (1) is proved. Inequality (2) follows from conclusion B of Lemma 6.

We call attention to the work of Hartman and Wintner who initiated the study of the positive temperatures on a finite rod [1950; 373]. Their results differ in appearance from those of Theorem 6, since they make no use of improper Stieltjes integrals.

7 EXAMPLES

We have seen that the basic functions $k(x, t)$ and $h(x, t)$, being $\geqslant 0$ in a quadrant, have the representation of Theorem 5.2. They also satisfy the hypotheses of Theorem 6. We determine here the specific integral transforms involved. As in §5, we give a proof by use of the product theorem for Laplace integrals. We derive here the Laplace transform of $\varphi(x, t)$ considered as a function of t.

Lemma 7

1. $0 < x < 2\pi$

$$\Rightarrow \int_0^\infty e^{-st}\varphi(x, t)\, dt = \frac{e^{(2\pi - x)\sqrt{s}} - e^{x\sqrt{s}}}{e^{2\pi\sqrt{s}} - 1}, \qquad 0 < s < \infty.$$

By Theorem 8.1, Chapter V,

$$\int_0^\infty e^{-st}\varphi(x, t)\, dt = \int_0^\infty e^{-st}\left[\sum_{n=0}^\infty h(x + 2n\pi, t)\, dt - \sum_{n=1}^\infty h(2n\pi - x, t)\right] dt$$

$$= e^{-x\sqrt{s}} + \sum_{n=1}^\infty e^{-(x+2n\pi)\sqrt{s}} - \sum_{n=1}^\infty e^{-(2n\pi - x)\sqrt{s}}$$

$$= e^{-x\sqrt{s}} - \frac{e^{x\sqrt{s}} - e^{-x\sqrt{s}}}{e^{2\pi\sqrt{s}} - 1} = \frac{e^{(2\pi - x)\sqrt{s}} - e^{x\sqrt{s}}}{e^{2\pi\sqrt{s}} - 1}$$

The term by term integration is justified since the integrands are ≥ 0; see Titchmarsh [1939; 44].

Theorem 7.1

$$1. \quad 0 < x < \pi, \quad 0 < t < \infty$$

$$\Rightarrow \quad h(x, t) = \varphi(x, t) + \int_0^t \varphi(\pi - x, t - y) h(\pi, y) \, dy. \tag{1}$$

We calculate the Laplace transform of the convolution (1). By the product theorem it is the transform of $h(\pi, y)$, which is $e^{-\pi\sqrt{s}}$, multiplied by the transform of $\varphi(\pi - x, y)$ given by Lemma 7:

$$e^{-\pi\sqrt{s}} \cdot \frac{e^{(\pi+x)\sqrt{s}} - e^{(\pi-x)\sqrt{s}}}{e^{2\pi\sqrt{s}} - 1} = \frac{e^{x\sqrt{s}} - e^{-x\sqrt{s}}}{e^{2\pi\sqrt{s}} - 1}. \tag{2}$$

But this is the transform of $h(x, t) - \varphi(x, t)$,

$$e^{-x\sqrt{s}} - \frac{e^{(2\pi-x)\sqrt{s}} - e^{x\sqrt{s}}}{e^{2\pi\sqrt{s}} - 1} = \frac{e^{x\sqrt{s}} - e^{-x\sqrt{s}}}{e^{2\pi\sqrt{s}} - 1}.$$

By the uniqueness theorem for Laplace transforms, equation (1) follows.

An alternative proof is provided by Corollary 10, Chapter V; for, by Theorem 5, Chapter V, $h(x, t) - \varphi(x, t) \in H$ in $-2\pi < x < 2\pi$. Hence

$$h(x, t) - \varphi(x, t) = \int_0^t \varphi(x, t - y)[h(0, y) - \varphi(0, y)] \, dy$$

$$+ \int_0^t \varphi(\pi - x, t - y)[h(\pi, y) - \varphi(\pi, y)] \, dy.$$

Since $[h(0, y) - \varphi(0, y)] = 0$ and $\varphi(\pi, y) = 0$, the right-hand side reduces to the convolution (1). Note that $h(x, t) - \varphi(x, t)$ has a removable singularity at $(0, 0)$.

Theorem 7.2

$$1. \quad 0 < x < \pi, \quad 0 < t < \infty$$

$$\Rightarrow \quad k(x, t) = \int_0^t \varphi(x, t - y) k(0, y) \, dy + \int_0^t \varphi(\pi - x, t - y) k(\pi, y) \, dy. \tag{3}$$

8. FURTHER CLASSES OF TEMPERATURE FUNCTIONS

Recalling that the Laplace transform of $k(x, t)$ is $e^{-x\sqrt{s}}/\sqrt{4s}$, we obtain the Laplace transform of the second convolution (3) by dividing equation (2) through by $\sqrt{4s}$. Hence we have only to show that

$$\frac{e^{-x\sqrt{s}}}{\sqrt{4s}} = \frac{1}{\sqrt{4s}} \left[\frac{e^{(2\pi - x)\sqrt{s}} - e^{x\sqrt{s}}}{e^{2\pi\sqrt{s}} - 1} \right] + \frac{e^{x\sqrt{s}} - e^{-x\sqrt{s}}}{\sqrt{4s}(e^{2\pi\sqrt{s}} - 1)}.$$

This is an obvious identity, so the proof is complete. The result could be predicted from the proof of Theorem 6. The first integral (3) is zero when $u(x, t) = k(x, t)$ so that the improper integral 6(1) disappears.

8 FURTHER CLASSES OF TEMPERATURE FUNCTIONS

We have thus far considered only bounded and positive solutions of the heat equation. It is possible to characterize other useful classes. Here we state a few of the results insofar as they concern the infinite rod, referring for proofs to Hirschman and Widder [1955, Chapter VIII].

Theorem 8.1

1. $u(x, t) \in H$, $\quad -\infty < x < \infty, \quad 0 < t < c,$
2. $\int_{-\infty}^{\infty} |u(x, t)|^p \, dx < M, \quad p > 1, \quad 0 < t < c,$

$\Leftrightarrow$
$$u(x, t) = \int_{-\infty}^{\infty} k(x - y, t) f(y) \, dy,$$

where

$$\int_{-\infty}^{\infty} |f(y)|^p \, dy < M.$$

Theorem 8.2

1. $u(x, t) \in H$, $\quad -\infty < x < \infty, \quad 0 < t < c,$
2. $\int_{-\infty}^{\infty} |u(x, t)| \, dx < M$

$\Leftrightarrow$
$$u(x, t) = \int_{-\infty}^{\infty} k(x - y, t) \, d\alpha(y),$$

where

$$\int_{-\infty}^{\infty} |d\alpha(y)| < M.$$

Corresponding results undoubtedly hold for the semi-infinite rod and for the finite rod. However, no reference is available.

Chapter IX

THE HUYGENS PROPERTY

1 INTRODUCTION

If the temperature of an infinite bar extended along the x-axis of an x, t-plane is given by the function $\varphi(x)$ at time $t = 0$, the temperature $u(x, t)$ at later times t is usually given by the Poisson integral

$$u(x, t) = e^{tD^2}\varphi(x) = \int_{-\infty}^{\infty} k(x - y, t)\varphi(y)\, dy, \qquad (1)$$

or at time $t + \delta$ by

$$u(x, t + \delta) = e^{(t+\delta)D^2}\varphi(x) = \int_{-\infty}^{\infty} k(x - y, t + \delta)\varphi(y)\, dy.$$

But if the integral (1) converges absolutely we have by substitution

$$\int_{-\infty}^{\infty} k(x - y, t)u(y, \delta)\, dy = \int_{-\infty}^{\infty} k(x - y, t)\, dy \int_{-\infty}^{\infty} k(y - z, \delta)\varphi(z)\, dz$$

$$= \int_{-\infty}^{\infty} \varphi(z)\, dz \int_{-\infty}^{\infty} k(x - y, t)k(y - z, \delta)\, dy$$

$$= \int_{-\infty}^{\infty} k(x - z, t + \delta)\varphi(z)\, dz = u(x, t + \delta). \qquad (2)$$

Here we have used Theorem 3, Chapter III. Symbolically, equation (2) is

$$e^{(t+\delta)D^2}\varphi(x) = e^{tD^2}(e^{\delta D^2}\varphi(x)).$$

Physically, equation (2) means that the temperature of the bar, $u(x, \delta)$, at time δ may be used as the data for Cauchy's problem to determine temperatures at later time. A similar phenomenon in optical theory was described by C. Huygens. The phenomenon does not obtain for all temperature functions, but when it does we shall describe it as the *Huygens property*. In Chapter II, §6, we saw the existence of "null" solutions of the heat equation, solutions $u(x, t)$ for which $u(x, 0) \equiv 0$ but which do not vanish identically for $t > 0$. Obviously, the Poisson integral (1) cannot be used to represent such a function for $t > 0$ with $\varphi(y) = u(y, 0)$. Accordingly, it becomes important to distinguish between those functions which have the Huygens property and those which do not.

Definition 1.1

$$u(x, t) \in H^o \quad \text{(Huygens property)}, \quad a < t < b,$$

$\Leftrightarrow$
A. $u(x, t) \in H, \quad a < t < b,$
B. $u(x, t) = \int_{-\infty}^{\infty} k(x - y, t - t')u(y, t') \, dy \quad (3)$

for every t and t', $a < t' < t < b$, the integral converging absolutely.

Let us introduce a further useful class of functions included in H^o, using the notation H^Δ introduced by Gehring [1960; 337].

Definition 1.2

1. $u(x, t) \in H^\Delta, \quad a < t < b,$

$\Leftrightarrow$ $u(x, t) = u_1(x, t) - u_2(x, t),$

where

$$u_1(x, t), u_2(x, t) \in H, \geqslant 0, \quad a < t < b.$$

The function $k(x, t)$ clearly belongs to both these classes in $0 < t < \infty$, for it is positive there and from Theorem 3, Chapter III,

$$k(x, t) = \int_{-\infty}^{\infty} k(x - y, t - t')k(y, t') \, dy, \quad 0 < t' < t < \infty. \quad (4)$$

In fact, we can show at once that $H^\Delta \subset H^o$.

1. INTRODUCTION

Theorem 1

$$\begin{aligned}&1. \quad u(x, t) \in H^\Delta, \qquad a < t < b, \\ \Rightarrow \quad & \quad u(x, t) \in H^\circ, \qquad a < t < b.\end{aligned}$$

It will be sufficient to assume $u(x, t) \geq 0$ and to take $a = 0$. Then by Theorem 3, Chapter VIII,

$$u(x, t) = \int_{-\infty}^{\infty} k(x - y, t)\, d\alpha(y), \qquad 0 < t < c,$$

where $\alpha(y) \in \uparrow(-\infty, \infty)$. To establish (3) we have

$$\int_{-\infty}^{\infty} k(x - y, t)\, d\alpha(y) = \int_{-\infty}^{\infty} k(x - y, t - t')\, dy \int_{-\infty}^{\infty} k(y - z, t')\, d\alpha(z)$$

$$= \int_{-\infty}^{\infty} d\alpha(z) \int_{-\infty}^{\infty} k(x - y, t - t') k(y - z, t')\, dy. \quad (5)$$

Here we have used Fubini's theorem to interchange the order of integration, applicable since $k > 0$ and $\alpha(z) \in \uparrow$. But the inner integral (5) is $k(x - z, t)$ by the addition formula, (4), and equation (3) is proved.

We point out that the classes H° and H^Δ are not identical. The function $h(x, t)$ provides an example of a temperature function in H° but not in H^Δ in the half-plane $0 < t < \infty$. For this function, (3) becomes

$$h(x, t) = k(x, t - t') * h(x, t'), \qquad 0 < t' < t < \infty. \quad (6)$$

To prove this we use the product theorem for bilateral Laplace transforms. Using Theorem 7.1 and Corollary 7.1, Chapter III, for the transforms of k and h, we must show that

$$-2se^{ts^2} = e^{(t-t')s^2}(-2se^{t's^2}).$$

But this is an obvious identity. That $h(x, t) \notin H^\Delta$ is not immediately evident. We defer the proof.

Corollary 1a

$$\begin{aligned}&1. \quad u(x, t) \in H^\circ, \qquad 0 < t < c, \\ &2. \quad 0 < \delta < c \\ \Rightarrow \quad & \quad u(x, t) \in H^\Delta, \qquad \delta < t < c.\end{aligned}$$

IX. THE HUYGENS PROPERTY

By Definition 1,

$$u(x, t) = \int_{-\infty}^{\infty} k(x - y, t - \delta) u(y, \delta) \, dy, \qquad \delta < t < c,$$

the integral converging absolutely. Replace $u(y, \delta)$ by the difference of two nonnegative functions:

$$u(x, t) = \int_{-\infty}^{\infty} k(x - y, t - \delta) \{ u(y, \delta) + |u(y, \delta)| \} \, dy$$

$$- \int_{-\infty}^{\infty} k(x - y, t - \delta) |u(y, \delta)| \, dy.$$

Each integral represents a function of H, ≥ 0, so that $u \in H^\Delta$.

Corollary 1b

1. $u(x, t) = \int_{-\infty}^{\infty} k(x - y, t) \, d\alpha(y)$ converges absolutely $\quad 0 < t < c$

$\Leftrightarrow \qquad u(x, t) \in H^\Delta, \qquad 0 < t < c.$

If we express $\alpha(y)$ as the difference of two nondecreasing functions [Widder, 1946; 6], it becomes clear that $u(x, t)$ is the difference of two functions of H, ≥ 0, and the sufficiency of condition 1 is proved. The necessity follows in an obvious way from Theorem 3, Chapter VIII.

Corollary 1c

1. $u(x, t) = \int_{-\infty}^{\infty} h(x - y, t) \, d\alpha(y)$ converges absolutely $\quad 0 < t < c$

$\Rightarrow \qquad u(x, t) \in H^o, \qquad 0 < t < c.$

Recall that $h(x, t)$ is an odd function of x so that we no longer have a positive kernel. By Definition 1.1 we must prove for $0 < t' < t < c$ that

$$\int_{-\infty}^{\infty} h(x - y, t) \, d\alpha(y) = \int_{-\infty}^{\infty} k(x - y, t - t') \, dy$$

$$\times \int_{-\infty}^{\infty} h(y - z, t') \, d\alpha(z). \qquad (7)$$

By Fubini's Theorem, applicable in the presence of hypothesis 1, the iterated integral (7) is equal to

$$\int_{-\infty}^{\infty} d\alpha(z) \int_{-\infty}^{\infty} k(x - y, t - t')h(y - z, t') \, dy$$

$$= \int_{-\infty}^{\infty} h(x - z, t) \, d\alpha(z).$$

Here we have evaluated the inner integral on the left by equation (6). Equation (7) is thus established.

2 BLACKMAN'S EXAMPLE

It is important to note that if $u(x, t) \in H^o$ in two adjoining or overlapping horizontal strips of the x, t-plane it need not do so in their union. The function

$$B(x, t) = k(x, t + i), \qquad i = \sqrt{-1},$$

is a case in point. The example was introduced by Blackman [1952; 678]. Hitherto we have dealt chiefly with real functions, but solutions of the heat equation may well be complex. If a real illustration is desired, the real part of B may be substituted for B in what follows. Explicitly,

$$B(x, t) = \frac{e^{-x^2/4(t+i)}}{\sqrt{4\pi(t+i)}},$$

where that branch of $\sqrt{t + ib}$ is chosen which is positive when $b = 0$, $t > 0$. The formal differentiation used to prove $k(x, t) \in H$ for $t > 0$ now shows that $B(x, t) \in H$ for $-\infty < t < \infty$. We prove the following result:

Theorem 2

1. $0 < a < \infty$

$\Rightarrow$ A. $B(x, t) \in H^o$, $-a < t < 1/a$,
           B. $B(x, t) \in H^o$, $0 < t < \infty$,
           C. $B(x, t) \in H^o$, $-\infty < t < 0$.

It will be sufficient to prove the first result since the other two are limiting cases, $a \to 0$ and $a \to \infty$. In Definition 1.1, choose $t' = -\delta$, $0 < \delta < a$. Then we must show that

$$B(x, t) = \int_{-\infty}^{\infty} k(x - y, t + \delta) B(y, -\delta) \, dy$$

$$= \int_{-\infty}^{\infty} k(x - y, t + \delta) k(y, i - \delta) \, dy \quad (1)$$

for $-\delta < t < 1/a$, the integral converging absolutely. Formally, this is just the addition formula for the source solution, Theorem 3, Chapter III, but we must reexamine its validity for complex values of the time variable. We prove the absolute convergence of the integral (1) first. The first factor of the integrand is real and positive. For the second,

$$|k(y, i - \delta)| = \frac{1}{\sqrt{4\pi(\delta^2 + 1)^{1/2}}} e^{\delta y^2/(4\delta^2 + 4)}.$$

The first factor predominates when

$$\frac{1}{t + \delta} > \frac{\delta}{\delta^2 + 1}, \quad (2)$$

or when $t < 1/\delta$, and hence when $t < 1/a$. That is, the integral (1) converges absolutely for $-\delta < t < 1/a$, as desired.

To prove the extended addition formula we use analytic continuation. Set

$$F(z) = \frac{1}{\sqrt{4\pi z}} \int_{-\infty}^{\infty} k(x - y, t + \delta) e^{-y^2/(4z)} \, dy, \quad z = a + ib, \quad (3)$$

so that the integral (1) is $F(i - \delta)$. If y^2 is replaced by r and z by $1/s$, the integral (3) becomes a Laplace integral, so that $F(1/s)$ is an analytic function of s in the half-plane of convergence of the Laplace integral [Widder, 1946; 57]. Hence, $F(z)$ is analytic for

$$\operatorname{Re} \frac{1}{z} > \frac{-1}{t + \delta}, \quad \frac{a}{a^2 + b^2} + \frac{1}{t + \delta} > 0.$$

This is a region D of the complex z-plane exterior to a disk with center at $z = -(t + \delta)/2$ and of radius $(t + \delta)/2$ (Figure 10). The positive real

3. CONDITIONALLY CONVERGENT POISSON INTEGRALS

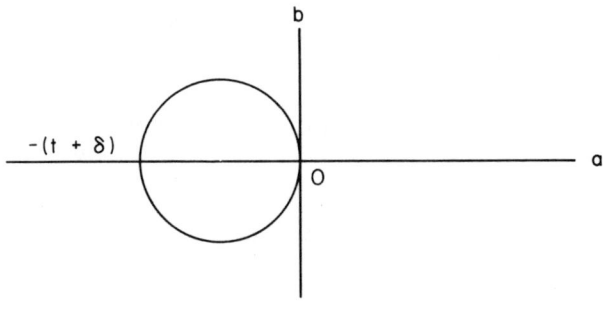

Figure 10

axis as well as the point in question, $z = -\delta + i$, are in D. But on the positive real axis,

$$F(a) = k(x, a + t + \delta), \quad a > 0,$$

by Theorem 3, Chapter III. But we see by inspection that $k(x, z + t + \delta)$ is analytic over the whole z-plane cut along the real axis from $-\infty$ to $-t - \delta$, and hence in D less the negative real axis. Since $F(z)$ and $k(x, z + t + \delta)$ coincide all along the whole positive axis of reals, they must coincide where both are analytic. Hence

$$F(-\delta + i) = k(x, t + i),$$

and this is equivalent to equation (1), which we wished to prove. Thus conclusion A is established.

Note that if $t > 1/\delta$ the reverse of inequality (2) holds and

$$\lim_{|y|\to\infty} k(x - y, t + \delta)|k(y, i - \delta)| = \infty,$$

so that the integral (1) fails to converge absolutely. That is, $B(x, t) \notin H^o$ in $-\delta < t < (1/\delta) + \epsilon$, $\epsilon > 0$. We can now substantiate the first statement in this section. For example, $B(x, t) \in H^o$ in $(-1, 1)$ and in $(-\frac{1}{2}, 2)$ but not in $(-1, 2)$.

3 CONDITIONALLY CONVERGENT POISSON INTEGRALS

We give first a simple example to show that Poisson integrals exist which converge but not absolutely in a strip of the x, t-plane. We then prove that an integration by parts replaces the given integral by an absolutely

IX. THE HUYGENS PROPERTY

convergent one. Finally, we show that any convergent Poisson integral defines a function of H^o. The example is

$$u(x, t) = \int_0^\infty k(x - y, t)\varphi(y)\, dy, \tag{1}$$

where

$$\varphi(y) = ye^{2y^2} \sin e^{y^2}.$$

Then

$$u(0, t) = \frac{1}{\sqrt{4\pi t}} \int_0^\infty e^{-y^2/4t} y e^{2y^2} \sin e^{y^2} dy$$

$$= \frac{1}{4\sqrt{\pi t}} \int_1^\infty \frac{\sin r}{r^p}\, dr, \qquad e^{y^2} = r; \tag{2}$$

$$p = \frac{1}{4t} - 1.$$

Since the familiar integral (2) is known to converge absolutely for $p > 1$ but only conditionally for $0 < p \leqslant 1$, we see that the Poisson integral (1) converges absolutely in the strip $0 < t < \frac{1}{8}$ but only conditionally in $\frac{1}{8} \leqslant t < \frac{1}{4}$. The question now arises: Does such an integral define a function having the Huygens property throughout its strip of convergence or only where it converges absolutely?

Theorem 3

1. $u(x, t) = \int_{-\infty}^\infty k(x - y, t)\, d\alpha(y)$ converges at (x_0, t_0)

$\Rightarrow \qquad u(x, t) \in H^o, \qquad 0 < t < t_0.$

Set

$$\beta(x) = \int_0^x k(x_0 - y, t_0)\, d\alpha(y),$$

so that, by hypothesis 1, $\beta(x)$ is bounded on $(-\infty, \infty)$ and

$$u(x, t) = \int_{-\infty}^\infty \frac{k(x - y, t)}{k(x_0 - y, t_0)}\, d\beta(y).$$

In the quotient of the integrand the dominant factor near $y = \pm\infty$ is $\exp((y^2/4t_0) - (y^2/4t))$ and this $\to 0$ as $|y| \to \infty$ for $0 < t < t_0$. Integrating by parts and using the fact that $k' = -h/2$, we obtain

$$u(x, t) = \frac{1}{2} \int_{-\infty}^{\infty} \beta(y) \frac{h(x_0 - y, t_0)}{k^2(x_0 - y, t_0)} k(x - y, t)\, dy$$

$$- \frac{1}{2} \int_{-\infty}^{\infty} \frac{\beta(y) h(x - y, t)\, dy}{k(x_0 - y, t_0)}.$$

Each integrand is

$$O(|y| e^{y^2(t-t_0)/(4tt_0)}), \qquad |y| \to \infty,$$

so that each integral converges absolutely for $0 < t < t_0$. Hence Corollaries 1b and 1c are applicable, and the proof is complete.

4 $H^o \neq H^\Delta$

In §1 we saw that class H^Δ is included in class H^o. We show now that the two classes are not identical. To do so we need only exhibit a function in H^o which is not in H^Δ. As indicated earlier, one such function is $h(x, t)$. In fact, any derivative of $k(x, t)$ is also such a function.

Theorem 4

$\Rightarrow$
1. $n = 1, 2, 3, \ldots$
   A. $\dfrac{\partial^n}{\partial x^n} k(x, t) \in H^o$, $\quad 0 < t < \infty$,
   B. $\dfrac{\partial^n}{\partial x^n} k(x, t) \notin H^\Delta$, $\quad 0 < t < \infty$.

It is easy to show by induction that

$$k^{(n)}(x, t) = \frac{k(x, t) P_n(x, t)}{t^n}, \tag{1}$$

where $P_n(x, t)$ is a polynomial of degree n for which

$$P_n(-x, t) = (-1)^n P_n(x, t). \tag{2}$$

For $n = 1$,

$$k'(x, t) = -\frac{x}{2t} k(x, t)$$

so that $P_1(x, t) = -x/2$. Similarly, $P_2(x, t) = (x^2 - 2t)/4$. Since $k(x, t) > 0$ for $t > 0$, $k(x, t) \in H^\circ$ and

$$k(x, t + \delta) = \int_{-\infty}^{\infty} k(x - y, t) k(y, \delta) \, dy, \qquad 0 < \delta < t + \delta,$$

$$= \int_{-\infty}^{\infty} k(y, t) k(x - y, \delta) \, dy.$$

By Chapter IV, §3, we may differentiate under the integral sign to obtain

$$k^{(n)}(x, t + \delta) = \int_{-\infty}^{\infty} k(y, t) k^{(n)}(x - y, \delta) \, dy$$

$$= \int_{-\infty}^{\infty} k(x - y, t) k^{(n)}(y, \delta) \, dy. \qquad (3)$$

Thus, $k^{(n)}(x, t)$ satisfies equation 1(3). To complete the proof of A we need only point out that the integral (3) converges absolutely. This is clear from equation (1). An alternative proof of (3) is provided by the product theorem for bilateral Laplace transforms, for,

$$s^n e^{ts^2} = \int_{-\infty}^{\infty} e^{-sy} k^{(n)}(y, t) \, dy, \qquad 0 < t < \infty, \qquad (4)$$

and the image of equation (3) is the obvious identity

$$s^n e^{(t+\delta)s^2} = e^{ts^2} s^n e^{\delta s^2}.$$

Equation (4) follows easily from

$$e^{ts^2} = \int_{-\infty}^{\infty} e^{-sy} k(y, t) \, dy$$

by successive integrations by parts.

To prove B we proceed by contradiction. Suppose $k^{(n)}(x, t) \in H^\Delta$, $0 < t < \infty$. Then by Theorem 3, Chapter VIII,

$$k^{(n)}(x, t) = \int_{-\infty}^{\infty} k(x - y, t) \, d\alpha(y), \qquad 0 < t < \infty, \qquad (5)$$

the integral converging absolutely. We may assume that $\alpha(y)$ is normalized,

$$\alpha(y) = \frac{\alpha(y+) + \alpha(y-)}{2}.$$

Then by Theorem 6, Chapter IV,

$$\alpha(x) - \alpha(0) = \lim_{t \to 0+} \int_0^x k^{(n)}(y, t) \, dy = k^{(n-1)}(x, 0+) - k^{(n-1)}(0, 0+). \tag{6}$$

Two cases arise according as n is even or odd. If n is even (2) shows that $k^{(n-1)}(0, t) = 0$ and

$$\alpha(x) - \alpha(0) = \lim_{t \to 0+} k^{(n-1)}(x, t) = 0, \qquad x \neq 0.$$

Here we have used (1) and Lemma 1, Chapter V. This equation shows that $\alpha(x)$ is constant and the integral (5) is identically zero, a contradiction.

If n is odd $k^{(n-1)}(x, 0+) = 0$, $x \neq 0$, and $|k^{(n-1)}(0, 0+)| = \infty$. Here equation (6) itself produces a contradiction, the left-hand side being finite, the right side infinite.

Note that equation (5) is possible if $n = 0$. The inversion formula becomes

$$\alpha(b) - \alpha(a) = \lim_{t \to 0+} \int_a^b k(y, t) \, dy = 1, \qquad a < 0 < b,$$

$$= 0, \qquad ab > 0,$$

so that $\alpha(t)$ is the step-function with a single unit jump at the origin.

Corollary 4

$$h(x, t) \notin H^\Delta, \qquad 0 < t < \infty.$$

This follows since $h(x, t) = -2k'(x, t)$.

5 HEAT POLYNOMIALS AND ASSOCIATED FUNCTIONS

In Chapter I, §5, we defined the heat polynomials $v_n(x, t)$ by the equation

$$e^{xz+tz^2} = \sum_{n=0}^{\infty} \frac{v_n(x, t) z^n}{n!}. \tag{1}$$

A function $w_n(x, t)$ associated with $v_n(x, t)$ is defined as the Appell transform thereof:

$$w_n(x, t) = \text{Ap}[v_n(x, t)] = k(x, t) v_n\left(\frac{x}{t}, -\frac{1}{t}\right), \qquad 0 < t < \infty. \quad (2)$$

We show here that $v_n(x, t) \in H^o$ over the whole x, t-plane and that $w_n(x, t) \in H^o$ for $0 < t < \infty$. Preliminary to the proof we need a crude upper bound for $|v_n(x, t)|$, but one that applies to all three variables, n, x, t.

Lemma 5

1. $-\infty < x < \infty, \qquad -\infty < t < \infty, \qquad n = 0, 1, 2, \ldots$

$\Rightarrow \qquad |v_n(x, t)| \leq n! \, e^{|x|+|t|}.$

By Cauchy's theorem applied to the power series (1)

$$v_n(x, t) = \frac{n!}{2\pi i} \int_\Gamma \frac{e^{xz+tz^2}}{z^{n+1}} \, dz,$$

where Γ is the unit circle

$$\Gamma = \{ z \mid z = e^{i\theta}, \; 0 \leq \theta < 2\pi \}.$$

On Γ

$$|xz + tz^2| \leq |xe^{i\theta} + te^{2i\theta}| \leq |x| + |t|,$$

so that

$$|v_n(x, t)| \leq n! \, e^{|x|+|t|},$$

as we wished to prove.

Theorem 5.1

1. $n = 0, 1, 2, \ldots$

$\Rightarrow \qquad v_n(x, t) \in H^o, \qquad -\infty < t < \infty.$

To prove this result, we must show for an arbitrary number a that

$$v_n(x, t) = \int_{-\infty}^{\infty} k(x - y, t + a) v_n(y, -a) \, dy, \qquad -a < t < \infty. \quad (3)$$

5. HEAT POLYNOMIALS AND ASSOCIATED FUNCTIONS

We prove first that this equation holds if $v_n(x, t)$ is replaced by its generating function (1)

$$e^{xz+tz^2} = \int_{-\infty}^{\infty} k(x-y, t+a)e^{yz-az^2}\, dy = e^{-az^2}e^{(t+a)D^2}e^{xz}. \quad (4)$$

But the right-hand side, by Theorem 7.1, Chapter III, is equal to

$$e^{-az^2}e^{xz+(t+a)z^2} = e^{xz+tz^2},$$

so that (4) is established. Substituting series (1) in (4), we have

$$\sum_{n=0}^{\infty} v_n(x, t) \frac{z^n}{n!} = \int_{-\infty}^{\infty} k(x-y, t+a) \sum_{n=0}^{\infty} v_n(y, -a) \frac{z^n}{n!}\, dy$$

$$= \sum_{n=0}^{\infty} \frac{z^n}{n!} \int_{-\infty}^{\infty} k(x-y, t+a) v_n(y, -a)\, dy \quad (5)$$

if term-by-term integration is valid. It is so if

$$\sum_{n=0}^{\infty} \frac{|z|^n}{n!} \int_{-\infty}^{\infty} k(x-y, t+a)|v_n(y, -a)|\, dy < \infty. \quad (6)$$

By Lemma 5, series (6) is dominated by

$$\sum_{n=0}^{\infty} \frac{|z|^n}{n!} n!\, e^{|a|} \int_{-\infty}^{\infty} k(x-y, t+a)e^{|y|}\, dy.$$

The integral converges for $t > -a$ and the series converges for $|z| < 1$. Hence equation (5) is valid, at least for $|z| < 1$. Comparing coefficients of z^n on two sides of the equation we obtain (3), and the theorem is proved.

Corollary 5.1

$$1.\quad 0 < t < \infty, \quad n = 0, 1, 2, \ldots.$$

$$\Rightarrow \quad v_n(x, t) = \int_{-\infty}^{\infty} k(x-y, t)y^n\, dy.$$

Since $v_n(y, 0) = y^n$, this is equation (3) with $a = 0$.

Theorem 5.2

$$\Rightarrow \quad \begin{array}{l} 1. \quad n = 0, 1, 2, \ldots \\ w_n(x, t) \in H^o, \quad 0 < t < \infty. \end{array}$$

If we replace z by λz in equation (1), we see at once that $v_n(x, t^2)$ is a homogeneous polynomial of degree n,

$$v_n(\lambda x, \lambda^2 t) = \lambda^n v_n(x, t).$$

Replacing t by $-t$ and then setting $\lambda = 1/t$, we obtain from (2)

$$w_n(x, t) = v_n(x, -t) t^{-n} k(x, t). \tag{7}$$

If we apply the Appell transform to both sides of equation (1), we obtain

$$\sum_{n=0}^{\infty} w_n(x, t) \frac{z^n}{n!} = \frac{e^{-x^2/(4t)}}{\sqrt{4\pi t}} e^{(xz - z^2)/t} = k(x - 2z, t). \tag{8}$$

Thus $k(x - 2z, t)$ is the generating function for the associated functions, just as $\exp(xz + tz^2)$ is for the heat polynomials. Comparing coefficients in series (8) with those of the Maclaurin expansion

$$k(x - 2z, t) = \sum_{n=0}^{\infty} \frac{k^{(n)}(x, t)(-2z)^n}{n!},$$

we obtain

$$w_n(x, t) = (-2)^n k^{(n)}(x, t), \quad n = 0, 1, 2, \ldots. \tag{9}$$

And now Theorem 5.2 is an immediate consequence of Theorem 4. Incidentally, equations (7) and (9) give an explicit determination of the polynomial $P_n(x, t)$ in equation 4(1). We have

$$k^{(n)}(x, t) = \frac{k(x, t) v_n(x, -t)}{(-2t)^n},$$

so that

$$P_n(x, t) = \frac{v_n(x, -t)}{(-2)^n}.$$

Chapter **X**

SERIES EXPANSIONS OF TEMPERATURE FUNCTIONS

1 INTRODUCTION

In this chapter we investigate the possibility of expanding a temperature function $u(x, t)$ in series of heat polynomials $v_n(x, t)$ or in series of associated functions $w_n(x, t)$:

$$u(x, t) = \sum_{n=0}^{\infty} a_n v_n(x, t), \qquad (1)$$

$$u(x, t) = \sum_{n=0}^{\infty} b_n w_n(x, t). \qquad (2)$$

From §5, Chapter IX we set down the generating functions for these two sets of functions:

$$e^{xz+tz^2} = \sum_{n=0}^{\infty} \frac{v_n(x, t)}{n!} z^n, \quad -\infty < t < \infty, \quad -\infty < z < \infty; \qquad (3)$$

$$k(x - 2z, t) = \sum_{n=0}^{\infty} \frac{w_n(x, t)}{n!} z^n, \quad 0 < t < \infty, \quad -\infty < z < \infty. \qquad (4)$$

X. SERIES EXPANSIONS OF TEMPERATURE FUNCTIONS

We recall that the Hermite polynomials, defined by

$$H_n(x) = (-1)^n e^{x^2} D^n e^{-x^2}, \quad n = 0, 1, 2, \ldots,$$

have a generating function similar to (3):

$$e^{2xz - z^2} = \sum_{n=0}^{\infty} \frac{H_n(x)}{n!} z^n.$$

Replacing x by $x/\sqrt{-4t}$ and z by $z\sqrt{-t}$, this equation becomes

$$e^{xz + tz^2} = \sum_{n=0}^{\infty} H_n\left(\frac{x}{\sqrt{-4t}}\right) (-t)^{n/2} \frac{z^n}{n!}. \tag{5}$$

Comparing coefficients of z^n in (3) and (5) we are able to express $v_n(x, t)$ in terms of $H_n(x)$ as follows:

$$v_n(x, t) = (-t)^{n/2} H_n\left(\frac{x}{\sqrt{-4t}}\right), \quad -\infty < x < \infty, \quad 0 < |t| < \infty. \tag{6}$$

From the familiar orthogonal relations among the Hermite polynomials we could establish a biorthogonal relation between the $v_n(x, t)$ and the $w_n(x, t)$. However, it is equally simple to prove it directly. We use the Kronecker delta, $\delta_{m,n} = 0, m \neq n, \delta_{m,m} = 1$.

Theorem 1

1. $0 < t < \infty, \quad m, n = 0, 1, 2, \ldots$

$$\Rightarrow \quad \int_{-\infty}^{\infty} v_n(x, -t) w_m(x, t) \, dx = \delta_{m,n} 2^n n! \, .$$

Note first that

$$v_n'(x, t) = n v_{n-1}(x, t).$$

This follows easily by differentiating equation (3) with respect to x and comparing coefficients in an obvious way. We need also equation 5(9), Chapter IX,

$$w_m(x, t) = (-2)^m k^{(m)}(x, t).$$

In what follows we use the letter A for an unessential constant that may vary from equation to equation. If $m > n$, we have by integration by parts, noting that $k^{(m)}(\pm \infty, t) = 0$,

$$\int_{-\infty}^{\infty} w_m(x, t) v_n(x, -t)\, dx = A \int_{-\infty}^{\infty} k^{(m)}(x, t) v_n(x, -t)\, dx$$

$$= A \int_{-\infty}^{\infty} k^{(m-n)}(x, t) v_0(x, -t)\, dx = 0.$$

If $m < n$,

$$\int_{-\infty}^{\infty} w_m(x, t) v_n(x, -t)\, dx = A \int_{-\infty}^{\infty} k(x, t) v_{n-m}(x, -t)\, dx$$

$$= A t^{n-m} \int_{-\infty}^{\infty} w_{n-m}(x, t)\, dx = 0.$$

In each case the final integrand is a derivative, which can then be integrated directly. If $m = n$

$$\int_{-\infty}^{\infty} w_n(x, t) v_n(x, -t)\, dx = (-2)^n \int_{-\infty}^{\infty} k^{(n)}(x, t) v_n(x, -t)\, dx$$

$$= 2^n n! \int_{-\infty}^{\infty} k(x, t) v_0(x, -t)\, dx = 2^n n! \, .$$

This completes the proof.

2 ASYMPTOTIC ESTIMATES

For a discussion of series 1(1) and 1(2) it is essential to know the behavior of $v_n(x, t)$ and $w_n(x, t)$ as $n \to \infty$.

Lemma 2.1

1. $0 < A, \quad 0 < t < 1/(4A)$

$$\Rightarrow \int_{-\infty}^{\infty} k(x - y, t) e^{Ay^2}\, dy = \frac{e^{Ax^2/(1-4At)}}{\sqrt{1 - 4At}}.$$

The change of variable $y - x = r$ gives

$$k * e^{Ax^2} = \frac{e^{Ax^2}}{\sqrt{4\pi t}} \int_{-\infty}^{\infty} \exp\left[-r^2(1 - 4At)(4t)^{-1} + 2Axr\right] dr.$$

Now complete the square in the exponent of the integrand and use the value of the probability integral

$$\int_{-\infty}^{\infty} e^{-r^2(1-4At)/(4t)} \, dr = \sqrt{\frac{4\pi t}{1-4At}}, \qquad 1 - 4At > 0.$$

Simple algebra then gives the desired result.

Theorem 2.1

1. $-\infty < x < \infty, \qquad 0 < t < \infty, \qquad \delta > 0, \qquad n = 1, 2, \ldots$

$$\Rightarrow \qquad |v_n(x, t)| \leq \left(\frac{2n}{e}\right)^{n/2} (t + \delta)^{(n+1)/2} \frac{e^{x^2/(4\delta)}}{\sqrt{\delta}}.$$

From Corollary 5.1, Chapter IX,

$$|v_n(x, t)| \leq \int_{-\infty}^{\infty} k(x - y, t)|y|^n \, dy. \tag{1}$$

By elementary calculus, we have for any constant $c > 0$

$$e^{-cy^2} y^n \leq \left(\frac{n}{2ec}\right)^{n/2}, \qquad 0 < y < \infty. \tag{2}$$

Taking $c = 1/[4(t + \delta)]$, we have

$$|y|^n \leq e^{y^2/4(t+\delta)} \left(\frac{2n(t + \delta)}{e}\right)^{n/2}.$$

Substituting in (1) and computing the integral by Lemma 2.1, we have

$$|v_n(x, t)| \leq \left(\frac{2n}{e}\right)^{n/2} \frac{(t + \delta)^{(n+1)/2}}{\sqrt{\delta}} e^{x^2/(4\delta)}.$$

This completes the proof. For negative t we use Cauchy's theorem as in the proof of Lemma 5, Chapter IX:

$$v_n(x, -t) = \frac{n!}{2\pi i} \int_\Gamma \frac{e^{xz-tz^2}}{z^{n+1}} \, dz, \qquad 0 < t,$$

$$\Gamma = \{z \mid z = re^{i\theta}, \ 0 \leq \theta < 2\pi\}.$$

2. ASYMPTOTIC ESTIMATES

Thus

$$|v_n(x, -t)| \leq \frac{n!}{r^n} \max_{0 \leq \theta \leq 2\pi} e^{xr \cos \theta - tr^2 \cos 2\theta}.$$

The function

$$f(\theta) = xr \cos \theta - tr^2 \cos 2\theta$$

has extremal values when $f'(\theta) = 0$ at $\theta = 0, \pi$ and when $\cos \theta = x/(4tr)$. For these values of θ, $f(\theta)$ is equal to

$$xr - tr^2, \quad -xr - tr^2, \quad tr^2 + \frac{x^2}{8t},$$

respectively. The first two values are less than the third since

$$(x \pm 4tr)^2 \geq 0.$$

Hence,

$$\max f(\theta) = tr^2 + \frac{x^2}{8t},$$

$$|v_n(x, -t)| \leq \frac{n!}{r^n} e^{tr^2 + x^2/(8t)}.$$

To make the most of this inequality we now choose r to minimize the right-hand side. By (2)

$$\frac{e^{tr^2}}{r^n} \geq \left(\frac{2et}{n} \right)^{n/2}.$$

Hence we have proved the following result:

Theorem 2.2

1. $-\infty < x < \infty, \quad 0 < t < \infty, \quad n = 1, 2, \ldots$

$$\Rightarrow \quad |v_n(x, -t)| \leq e^{x^2/(8t)} \left(\frac{2et}{n} \right)^{n/2} n!.$$

Corollary 2.2

1. $0 < t < \infty$

$$\Rightarrow \quad |w_n(x, t)| \leq \frac{e^{-x^2/(8t)}}{\sqrt{4\pi t}} \left(\frac{2e}{nt} \right)^{n/2} n!.$$

This follows from the relation

$$w_n(x, t) = \frac{k(x, t)v_n(x, -t)}{t^n}.$$

For most of our results the upper bounds thus far obtained are quite adequate. However, at least one subsequent theorem will require a more delicate estimate. But this is available from known facts about Hermite polynomials. From Erdèlyi [1953; 201] we have

$$\Gamma\left(\frac{n}{2} + 1\right) e^{-x^2/2} \frac{H_n(x)}{n!} = \cos\left(x\sqrt{2n+1} - \frac{n\pi}{2}\right)$$

$$+ O\left(\frac{1}{\sqrt{n}}\right), \qquad n \to \infty,$$

where the term $O(1/\sqrt{n})$ holds uniformly over any finite interval $|x| \leq R$. This result, by use of equation 1(5), gives

$$v_n(x, -t) = t^{n/2} H_n\left(\frac{x}{\sqrt{4t}}\right)$$

$$= \frac{n!}{\Gamma((n/2) + 1)} t^{n/2} e^{x^2/(8t)} \left[\cos\left(\frac{x\sqrt{2n+1}}{\sqrt{4t}} - \frac{n\pi}{2}\right) + O\left(\frac{1}{\sqrt{n}}\right)\right].$$

By Stirling's formula,

$$\frac{n!}{\Gamma((n/2)+1)} \sim \sqrt{2}\left(\frac{2n}{e}\right)^{n/2}, \qquad n \to \infty.$$

We thus obtain the following result:

Theorem 2.3

$$1. \quad r > 0, \qquad R > 0, \qquad t > 0$$

$$\Rightarrow v_n(x, -t) = \sqrt{2}\, e^{x^2/(8t)} \left(\frac{2nt}{e}\right)^{n/2}$$

$$\times \left\{\cos\left[x(2n+1)^{1/2}(4t)^{-1/2} - n\pi/2\right] + O(n^{-1/2})\right\},$$

3 A GENERATING FUNCTION

where $O(n^{-1/2})$ holds uniformly in $|x| \leq R$, $t \geq r$.

3 A GENERATING FUNCTION

We now develop a generating function for the biorthogonal set $v_n(x, t)$, $w_n(y, s)$.

Lemma 3

1. $-\infty < x < \infty$, $\quad 0 < t < \infty$

$\Rightarrow \quad \int_{-\infty}^{\infty} k(r + ix, t)(ir)^n \, dr = v_n(x, -t).$

From the Laplace transform of $k(x, t)$, Theorem 7.1, Chapter III, we have

$$\int_{-\infty}^{\infty} e^{irz} k(r, t) \, dr = e^{-tz^2},$$

or, by change of variable,

$$\int_{-\infty}^{\infty} e^{irz} k(r + ix, t) \, dr = e^{xz - tz^2}.$$

Expanding in powers of z

$$\sum_{n=0}^{\infty} \frac{z^n}{n!} \int_{-\infty}^{\infty} k(r + ix, t)(ir)^n dr = \sum_{n=0}^{\infty} v_n(x, -t) \frac{z^n}{n!}.$$

The term-by-term integration is easily justified for $t > 0$ and all z. The proof is completed by comparing coefficients of z^n.

Theorem 3

1. $-\infty < x < \infty$, $\quad -\infty < y < \infty$, $\quad -s < t < s$

$\Rightarrow \quad k(x - y, s + t) = \sum_{n=0}^{\infty} \frac{v_n(x, t) w_n(y, s)}{n! \, 2^n}.$ \quad (1)

From equation 1(4) we have

$$k(y - r, s) = \sum_{n=0}^{\infty} \frac{r^n w_n(y, s)}{2^n n!}.$$

Multiply both sides by $k(x - r, t)$ and integrate with respect to r. Since $k(x, s) \in H^o$ for $s > 0$ the left-hand side becomes $k(y - r, s + t)$ and since $k(x, t) * x^n$ is $v_n(x, t)$ we obtain (1) for $t > 0$ if term-by-term integration is valid. It will be so if

$$\sum_{n=0}^{\infty} \frac{|w_n(y, s)|}{n! \, 2^n} \int_{-\infty}^{\infty} k(x - r, t)|r|^n \, dr < \infty. \tag{2}$$

The integral (2) is the same as 1(1) used in the proof of Theorem 2.1, so that we may use the conclusion of that theorem. That result and Corollary 2.2 gives the following dominant series for (2):

$$\sum_{n=0}^{\infty} \frac{e^{-y^2/(8s)}}{\sqrt{4\pi s}} \frac{1}{2^n} \left(\frac{2e}{ns} \right)^{n/2} \frac{e^{x^2/(4\delta)}}{\sqrt{\delta}} \left(\frac{2n}{e} \right)^{n/2} (t + \delta)^{(n+1)/2}.$$

This reduces, except for unessential factors, to the geometric series

$$\sum_{n=0}^{\infty} \left(\frac{t + \delta}{s} \right)^{n/2},$$

which converges for $t + \delta < s$. Since δ is arbitrary we have proved (1) for $0 < t < s$.

For negative t we use Lemma 3. Equation 1(4) gives

$$k(y - ir, s) = \sum_{n=0}^{\infty} \frac{(ir)^n}{2^n n!} w_n(y, s), \qquad 0 < s.$$

Multiplying by $k(r + ix, t)$ and integrating with respect to r, we have

$$k(x - y, s - t) = \sum_{n=0}^{\infty} \frac{w_n(y, s)}{2^n n!} \int_{n=0}^{\infty} k(r + ix, t)(ir)^n \, dr.$$

By Lemma 3 this is equation (1), and we need only justify the term-by-term integration, valid if

$$\sum_{n=0}^{\infty} \frac{|w_n(y, s)|}{2^n n!} \int_{-\infty}^{\infty} |k(r + ix, t)| \, |r|^n \, dr < \infty$$

or

$$e^{x^2/(4t)} \sum_{n=0}^{\infty} \frac{|w_n(y, s)|}{2^n n!} \int_{-\infty}^{\infty} k(r, t)|r|^n \, dr < \infty.$$

But this is true by (2), $x = 0$, for $0 < t < s$. That is, (1) is valid for $-s < t < s$, as we wished to prove.

4 REGION OF CONVERGENCE

We show here that if series 1(1) converges at a point (x_0, t_0) with $t_0 > 0$ and $x_0 \neq 0$, then it converges absolutely in the strip $|t| < t_0$.

Theorem 4.1

1. $\sum_{n=0}^{\infty} a_n v_n(x_0, t_0)$ converges, $t_0 > 0$, $x_0 \neq 0$

$\Rightarrow \quad \sum_{n=0}^{\infty} a_n v_n(x, t)$ converges absolutely, $|t| < t_0$.

Since the coefficients of $v_n(x, t)$ are all > 0, we have for $t > 0$

$$v_{2n}(x, t) \geq v_{2n}(0, t) = \frac{(2n)! \, t^n}{n!}. \tag{1}$$

Similarly, $|v_{2n+1}(x, t)/x| \geq$ its value at $x = 0$, which we compute by L'Hospital's rule:

$$|v_{2n+1}(x, t)| \geq |x| v'_{2n+1}(0, t) = |x|(2n + 1) v_{2n}(0, t) = \frac{|x|(2n + 1)! \, t^n}{n!}. \tag{2}$$

By hypothesis 1, the general term $a_n v_n(x_0, t_0)$ remains bounded as $n \to \infty$. Hence by (1) and (2),

$$a_{2n} = O\left(\frac{n!}{(2n)! \, t_0^n}\right) = O\left(\left[\frac{e}{4nt_0}\right]^n\right), \quad n \to \infty, \tag{3}$$

$$a_{2n+1} = O\left(\frac{n!}{(2n + 1)! \, t_0^n}\right) = O\left(\left[\frac{e}{(4n + 2)t_0}\right]^{(2n+1)/2}\right), \quad n \to \infty. \tag{4}$$

Here we have used Stirling's formula

$$n! \sim \left(\frac{n}{e}\right)^n \sqrt{2\pi n}, \quad n \to \infty.$$

Relations (3) and (4) are equivalent to

$$a_n = o\left(\frac{e}{2nt_0}\right)^{n/2}, \quad n \to \infty.$$

Now by use of Theorem 2.1 we have for $t > 0$ and some M independent of n,

$$\sum_{n=0}^{\infty} a_n v_n(x, t) \ll M \sum_{n=0}^{\infty} \left(\frac{e}{2nt_0}\right)^{n/2} \left(\frac{2n}{e}\right)^{n/2} (t + \delta)^{(n+1)/2},$$

and by Theorem 2.2 for $t < 0$

$$\sum_{n=0}^{\infty} a_n v_n(x, t) \ll M \sum_{n=0}^{\infty} \left(\frac{e}{2nt_0}\right)^{n/2} \left(\frac{2e|t|}{n}\right)^{n/2} n!.$$

Applying Sterling's formula in the latter series we see that in each case the dominant series is essentially a geometric series, the first converging for $t + \delta < t_0$, the second for $|t| < t_0$. Since δ is arbitrary the theorem is proved.

Thus convergence of our series at (x_0, t_0), $t_0 > 0$, implies its convergence in a strip symmetric about the x-axis. Is there a corresponding result when $t_0 < 0$? Here we need to postulate convergence over an interval rather than at a point.

Theorem 4.2

1. $\sum_{n=0}^{\infty} a_n v_n(x, -t_0)$ converges, $\quad a < x < b, \quad t_0 > 0 \quad (5)$

$$\Rightarrow \quad a_n = o\left(\frac{e}{2nt_0}\right)^{n/2}, \quad n \to \infty.$$

Suppose, on the contrary, that $a_n c_n \not\to 0$, where

$$c_n = \left(\frac{2nt_0}{e}\right)^{n/2}.$$

4. REGION OF CONVERGENCE

Then there would exist a constant A and infinitely many integers m such that

$$|a_m c_m| > A > 0.$$

Now the general term of series (5) $\to 0$, so that

$$\lim_{m \to \infty} \frac{a_m^2 v_m^2(x, -t_0)}{a_m^2 c_m^2} = 0, \qquad a < x < b. \tag{6}$$

But the quotient (6) is uniformly bounded on $a < x < b$; for, we have by Theorem 2.3

$$\left| \frac{v_m(x, -t_0)}{c_m} - \sqrt{2}\, e^{x^2/(8t_0)} \cos\left[x(2m+1)^{1/2}(4t_0)^{-1/2} - \frac{m\pi}{2} \right] \right|$$

$$\leq \frac{M}{\sqrt{m}}, \qquad a < x < b, \tag{7}$$

where M is a constant independent of x and m. Thus, if N is an upper bound of the modulus of the second term in (7) on $a < x < b$, we have

$$\left| \frac{v_m(x, -t_0)}{c_m} \right| \leq M + N, \qquad a < x < b.$$

We may consequently apply Lebesgue's limit theorem to show that

$$\lim_{m \to \infty} \int_a^b \frac{v_m^2(x, -t_0)}{c_m^2}\, dx = 0. \tag{8}$$

Write (7) as

$$\frac{v_m(x, -t_0)}{c_m} = \sqrt{2}\, e^{x^2/(8t_0)} \cos\left(x\sqrt{\frac{2m+1}{4t_0}} - \frac{m\pi}{2} \right) + O(m^{-1/2})$$

and square both sides to obtain

$$\left(\frac{v_m(x, -t_0)}{c_m} \right)^2 = 2 e^{x^2/(4t_0)} \cos^2\left(x\sqrt{\frac{2m+1}{4t_0}} - \frac{m\pi}{2} \right) + O(m^{-1/2}),$$

the term $O(1/m)$ being negligible. Integrate both sides of this equation and apply (8):

$$\lim_{m\to\infty} \int_a^b e^{x^2/(4t_0)} \cos^2\left(x\sqrt{\frac{2m+1}{4t_0}} - \frac{m\pi}{2}\right) dx = 0,$$

$$\int_a^b e^{x^2/(4t_0)} dx \pm \lim_{m\to\infty} \int_a^b e^{x^2/(4t_0)} \cos\left(x\sqrt{\frac{2m+1}{t_0}}\right) dx = 0.$$

Here we have used trigonometric half-angle formulas. The second term on the left is zero by the Riemann–Lebesgue theorem [Widder, 1961c; 404]. Hence we reach the evident contradiction that the first integral, with a positive integrand, is zero. We must conclude that $a_n c_n \to 0$, as we wished to prove.

Corollary 4.2 Series (5) converges absolutely, $|t| < t_0$.

5 STRIP OF CONVERGENCE

There exist series

$$\sum_{n=0}^{\infty} a_n v_n(x, t) \tag{1}$$

which converge for all (x, t), others which fail to converge when $t \neq 0$. Putting aside these cases and imitating an argument familiar for power series we see from Theorem 4.1 and Corollary 4.2 that there must exist a number σ such that (1) converges for $|t| < \sigma$ and fails to do so throughout the strip $|t| < \sigma + \epsilon$, $\epsilon > 0$. We define this number σ as the *radius of convergence* and the strip $|t| < \sigma$ as the *strip of convergence* for the series (1). The above limiting cases correspond to $\sigma = \infty$ and $\sigma = 0$, respectively. We derive now a formula for σ in terms of the coefficients a_n, analogous to Hadamard's formula for the radius of convergence for power series.

Theorem 5.1

$$1. \quad \overline{\lim_{n\to\infty}} \frac{|a_n|^{2/n}(2n)}{e} = \frac{1}{\sigma}$$

$\Rightarrow$ (1) converges absolutely for $|t| < \sigma$ and does not converge everywhere for $|t| < \sigma + \epsilon$, $\epsilon > 0$.

5. STRIP OF CONVERGENCE

Obvious modifications of the wording are needed for the limiting cases $\sigma = 0$ or $\sigma = \infty$, but the formula remains accurate. We give the proof when $0 < \sigma < \infty$. By hypothesis 1,

$$a_n = O\left(\frac{e}{2n\sigma'}\right)^{n/2}, \quad n \to \infty,$$

for every $\sigma' < \sigma$. By Theorem 2.1, we have for $0 < t$

$$\sum_{n=0}^{\infty} a_n v_n(x, t) \ll M \sum_{n=0}^{\infty} \left(\frac{e}{2n\sigma'}\right)^{n/2} \left(\frac{2n}{e}\right)^{n/2} (t + \delta)^{(n+1)/2} \frac{e^{x^2/(4\delta)}}{\sqrt{\delta}}.$$

The dominant series converges for $t + \delta < \sigma'$. Hence (1) converges absolutely for $0 < t < \sigma$. By Theorem 2.2,

$$\sum_{n=0}^{\infty} a_n v_n(x, -t) \ll M \sum_{n=0}^{\infty} \left(\frac{e}{2n\sigma'}\right)^{n/2} \left(\frac{2et}{n}\right)^{n/2} n! \, e^{x^2/(8t)},$$

and again the dominant series converges for $0 < t < \sigma'$. Consequently (1) converges absolutely for $-\sigma < t < \sigma$.

To prove the second part, assume that (1) converged on the line $t = -t' < -\sigma$. By Theorem 4.2,

$$a_n = O\left(\frac{e}{2nt'}\right)^{n/2}, \quad n \to \infty,$$

whence

$$\overline{\lim_{n \to \infty}} \frac{|a_n|^{2/n}(2n)}{e} \leqslant \frac{1}{t'}$$

or

$$\frac{1}{\sigma} \leqslant \frac{1}{t'} < \frac{1}{\sigma}.$$

The desired contradiction is evident.

It is to be noted that the theorem does not preclude the possibility of the convergence of (1) outside the strip of convergence. This may happen, as in the example

$$u(x, t) = e^{tD^2} x e^{x^2} = \sum_{n=0}^{\infty} \frac{v_{2n+1}(x, t)}{n!}. \tag{2}$$

Here

$$\lim_{n\to\infty} \frac{|a_n|^{2/n}(2n)}{e} = 4.$$

The strip of convergence is $|t| < \frac{1}{4}$, but the series (2) converges over the whole t-axis, where the general term is zero. An explicit form for the function (2) may be obtained from Lemma 2.1:

$$u(x, t) = x(1 - 4t)^{-3/2} e^{x^2/(1-4t)}, \quad 0 < t < \tfrac{1}{4}.$$

As direct confirmation of the width of the strip of convergence we may cancel a factor x from both sides of (2) and set $x = 0$ to obtain

$$(1 - 4t)^{-3/2} = \sum_{n=0}^{\infty} \frac{v'_{2n+1}(0, t)}{n!} = \sum_{n=0}^{\infty} \frac{(2n + 1)!}{n!\, n!} t^n.$$

This is a familiar binomial series whose radius of convergence is $\tfrac{1}{4}$.

A formula for computing the region of convergence of a series of associated functions follows.

Theorem 5.2

1. $\overline{\lim_{n\to\infty}} \dfrac{|b_n|^{2/n}(2n)}{e} = \sigma < \infty$

$\Rightarrow \quad \displaystyle\sum_{n=0}^{\infty} b_n w_n(x, t) \quad$ converges absolutely for $\quad t > \sigma$

and does not converge everywhere for $t > \sigma - \epsilon$, $\epsilon > 0$.
Series (3) is equal to

$$k(x, t) \sum_{n=0}^{\infty} b_n v_n\!\left(\frac{x}{t}, -\frac{1}{t}\right), \quad 0 < t,$$

and, by Theorem 5.1, it converges absolutely for $-1/\sigma < -1/t$ and does not converge everywhere for $-1/(\sigma - \epsilon) < -1/t$. But this is equivalent to the statement of the theorem.

As an example, we may apply the Appell transformation to equation (2) to obtain

$$\frac{x e^{-x^2/4(t+4)}}{\sqrt{4\pi(t + 4)^3}} = \sum_{n=0}^{\infty} \frac{w_{2n+1}(x, t)}{n!}.$$

Again checking at $x = 0$ we find

$$\frac{1}{\sqrt{(t+4)^3}} = \frac{1}{\sqrt{t^3}} \sum_{n=0}^{\infty} \frac{(2n+1)!}{n!\,n!} \left(\frac{-1}{t}\right)^n, \qquad t > 4.$$

6 REPRESENTATION BY SERIES OF HEAT POLYNOMIALS

Analogous to the classic result concerning representation of analytic functions by power series we now prove that a temperature function can be expanded in series of heat polynomials in the widest strip $|t| < \sigma$ in which it enjoys the Huygens property.

Theorem 6

1. $u(x, t) \in H^o$, $\qquad |t| < \sigma$,

$\Leftrightarrow \qquad u(x, t) = \sum_{n=0}^{\infty} a_n v_n(x, t), \qquad |t| < \sigma,$ (1)

where

$$a_n = \frac{u^{(n)}(0, 0)}{n!} \qquad (2)$$

$$= \frac{1}{2^n n!} \int_{-\infty}^{\infty} u(y, -t) w_n(y, t)\, dy, \qquad 0 < t < \sigma. \qquad (3)$$

First assume equation (1) and show that $u(x, t) \in H^o$. We must show that for every c, $0 < c < \sigma$,

$$u(x, t) = \int_{-\infty}^{\infty} k(x - y, t + c) u(y, -c)\, dy, \qquad |t| < c, \qquad (4)$$

the integral converging absolutely. Substituting series (1) in integral (4) and recalling that $v_n(x, t) \in H^o$, $-\infty < t < \infty$, we have

$$u(x, t) = \sum_{n=0}^{\infty} a_n \int_{-\infty}^{\infty} k(x - y, t + c) v_n(y, -c)\, dy$$

$$= \sum_{n=0}^{\infty} a_n v_n(x, t),$$

provided term-by-term integration is valid. It will be so if

$$\sum_{n=0}^{\infty} |a_n| \int_{-\infty}^{\infty} k(x - y, t + c)|v_n(y, - c)| \, dy < \infty. \tag{5}$$

By Theorem 4.1,

$$a_n = O\left(\frac{e}{2nt'}\right)^{n/2}, \quad n \to \infty,$$

for any $t' < \sigma$, and, by Theorem 2.2,

$$|v_n(x, - c)| \leq e^{x^2/(8c)}(2ec/n)^{n/2} n! \ .$$

Hence series (5) is dominated by

$$\int_{-\infty}^{\infty} k(x - y, t + c)e^{y^2/(8c)} \, dy \, M \sum_{n=0}^{\infty} \left(\frac{e}{2nt'}\right)^{n/2} \left(\frac{2ec}{n}\right)^{n/2} n! \ .$$

The integral converges for $|t| < c$ since $0 < 1/(t + c) > 1/(2c)$ there. The series converges for $t' > c$ by the ratio test. Hence, if we choose t' so that $c < t' < \sigma$, we see that (5) is true for $|t| < c$ and (4) is established.

Conversely, assume $u(x, t) \in H^o$ in $|t| < \sigma$. For every such t

$$u(x, t) = \int_{-\infty}^{\infty} k(x - y, t + s)u(y, - s) \, dy, \tag{6}$$

where $-\sigma < -s < t \leq s < \sigma$ and the integral converges absolutely, by the definition of H^o. By Theorem 3

$$k(x - y, s + t) = \sum_{n=0}^{\infty} \frac{v_n(x, t)w_n(y, s)}{n! \, 2^n}, \quad |t| < s.$$

Substituting this series in (6) we obtain

$$u(x, t) = \sum_{n=0}^{\infty} a_n v_n(x, t),$$

$$a_n = \frac{1}{2^n n!} \int_{-\infty}^{\infty} u(y, - s)w_n(y, s) \, dy,$$

provided term-by-term integration is valid. It will be so if

$$\sum_{n=0}^{\infty} \frac{|v_n(x, t)|}{n! \, 2^n} \int_{-\infty}^{\infty} |u(y, - s)| \, |w_n(y, s)| \, dy < \infty. \tag{7}$$

But by Theorem 2.1 and Corollary 2.2, series (7) is dominated when $t > 0$ by

$$\frac{e^{x^2/(4\delta)}}{\sqrt{4\pi s\delta}} \sum_{n=0}^{\infty} \frac{1}{n! \, 2^n} \left(\frac{2n}{e}\right)^{n/2} (t+\delta)^{(n+1)/2}$$

$$\times \left(\frac{2e}{ns}\right)^{n/2} n! \int_{-\infty}^{\infty} e^{-y^2/(8s)} |u(y, -s)| \, dy. \tag{8}$$

The integral (8) converges, as we see by setting $x = 0$, $t = s$ in (6). The series converges for $t + \delta < s$, by the ratio test. When $t < 0$, we use Theorem 2.2 instead of 2.1 and conclude that (8) converges for $-s < t < 0$. Hence (1) is established with the integral determination (3) of the coefficients a_n. To obtain the derivative form (2) set $t = 0$ in (1):

$$u(x, 0) = \sum_{n=0}^{\infty} a_n x^n, \quad -\infty < x < \infty.$$

Then equation (2) follows from Taylor's formula. Despite appearances integral (3) is independent of t, for, by the above proof,

$$\int_{-\infty}^{\infty} u(y, -t) w_n(y, t) \, dy = 2^n u^{(n)}(0, 0), \quad 0 < t < \sigma.$$

This could be proved directly. See Widder [1959; 246].

7 THE GROWTH OF AN ENTIRE FUNCTION

In later work we shall need certain known facts about entire functions. We record them here and refer for proofs to Boas [1954; 11], for example. An entire function is said to have *growth* $\{\rho, \sigma\}$ if and only if it has order $\leq \rho$, and if the order $= \rho$ then the type is $\leq \sigma$. We need only $\rho = 2$, so we restrict our formal definition to that case.

Definition 7

$$f(z) \in \{2, \sigma\}, \quad \text{has growth} \quad \{2, \sigma\},$$

$\Leftrightarrow \quad f(z)$ is entire of order < 2 or is of order 2 and of type $\leq \sigma$.

Referring to the definitions of order and type, we see that an entire function $\in \{2, \sigma\}$ if and only if, for every $\epsilon > 0$,

$$f(z) = O(e^{(\sigma+\epsilon)|z|^2}), \quad |z| \to \infty.$$

If

$$f(z) = \sum_{n=0}^{\infty} a_n z^n, \qquad (1)$$

another criterion is available in terms of the coefficients a_n:

$$\varlimsup_{n \to \infty} n|a_n|^{2/n} \leq 2e\sigma. \qquad (2)$$

Since this formula involves only absolute values, the function

$$\sum_{n=0}^{\infty} |a_n| z^n \qquad (3)$$

belongs to the same class as (1). For real y, $|y| = y$, $y > 0$; $|y| = -y$, $y < 0$. Hence if the function (1) $\in \{2, \sigma\}$, then for real y and every $\epsilon > 0$,

$$\sum_{n=0}^{\infty} |a_n| |y|^n = O(e^{(\sigma+\epsilon)y^2}), \qquad |y| \to \infty. \qquad (4)$$

We shall need this fact in §8.

As an example, $h(z, t) \in \{2, 1/(4t)\}$, $t \neq 0$. Note that $e^{az} \in \{2, \sigma\}$ for any σ since its order is $1 < 2$. Of course it also belongs to $\{1, |a|\}$.

8 EXPANSIONS IN SERIES OF ASSOCIATED FUNCTIONS

We give two criteria for the expansibility of a temperature function in terms of the $w_n(x, t)$.

Theorem 8

$$1. \quad u(x, t) = \sum_{n=0}^{\infty} b_n w_n(x, t), \qquad t > \sigma \geq 0, \qquad (1)$$

$$\Leftrightarrow \quad u(x, t) = \frac{1}{2\pi} \int_{-\infty}^{\infty} e^{ixy - ty^2} \varphi(y) \, dy, \qquad \varphi(y) \in \{2, \sigma\}. \qquad (2)$$

By Theorem 5, Chapter III,

$$k(x, t) = \frac{1}{2\pi} \int_{-\infty}^{\infty} e^{ixy - ty^2} \, dy, \qquad 0 < t,$$

8. EXPANSIONS IN SERIES OF ASSOCIATED FUNCTIONS

and by 5(9), Chapter IX,

$$w_n(x, t) = (-2)^n k^{(n)}(x, t) = \int_{-\infty}^{\infty} e^{ixy - ty^2}(-2iy)^n \, dy.$$

If

$$\varphi(y) = \sum_{n=0}^{\infty} a_n y^n,$$

series (1) will be equal to integral (2) if and only if

$$\frac{1}{2\pi} \int_{-\infty}^{\infty} e^{ixy - ty^2} \sum_{n=0}^{\infty} a_n y^n \, dy = \sum_{n=0}^{\infty} \frac{a_n}{(-2i)^n} w_n(x, t).$$

The term-by-term integration will be valid if

$$\int_{-\infty}^{\infty} e^{-ty^2} \sum_{n=0}^{\infty} |a_n| \, |y|^n \, dy < \infty. \tag{3}$$

Now for the sufficiency of the condition (2) we are assuming that $\varphi(y) \in \{2, \sigma\}$, so that relation 7(4) holds. That is, integral (3) is dominated by

$$M \int_{-\infty}^{\infty} e^{-ty^2} e^{(\sigma + \epsilon)y^2} \, dy, \quad \text{some } M.$$

It converges for $t > \sigma + \epsilon$, and hence (1) holds for $t > \sigma$, and the sufficiency is proved.

Conversely, if (1) converges for $t > \sigma$, it converges on the whole line $t = \sigma + \epsilon$, $\epsilon > 0$. By 5(7), Chapter IX,

$$\sum_{n=0}^{\infty} b_n(\sigma + \epsilon)^{-n} v_n(x, -\sigma - \epsilon)$$

converges for all x. Hence by Theorem 4.2,

$$\frac{b_n}{(\sigma + \epsilon)^n} = O\left(\frac{e}{2n(\sigma + \epsilon)}\right)^{n/2}, \quad n \to \infty.$$

If $a_n = (-2i)^n b_n$, then

$$a_n = O\left(\frac{2e(\sigma + \epsilon)}{n}\right)^{n/2},$$

or

$$\varlimsup_{n\to\infty} \frac{n}{2e} |a_n|^{2/n} \leq \sigma + \epsilon.$$

That is, by 7(2), $\varphi \in \{2, \sigma + \epsilon\}$ for every $\epsilon > 0$, or $\varphi \in \{2, \sigma\}$. Then (3) follows as before and the proof is complete.

Corollary 8

1. $u(x, t) = \sum_{n=0}^{\infty} b_n w_n(x, t), \qquad t > \sigma \geq 0,$

$\Rightarrow \qquad b_n = \dfrac{\varphi^{(n)}(0)}{n!(-2i)^n}.$

This results from Maclaurin's theorem.

9 A FURTHER CRITERION

Theorem 8 has the disadvantage that the test for the expansibility of $u(x, t)$ does not bear on $u(x, t)$ itself but rather on another related function $\varphi(y)$ which may not be easily found. Here we develop a second criterion which does not have that disadvantage. We need a preliminary result, of interest in itself.

Theorem 9.1

1. $u(x, t) = \dfrac{1}{2\pi} \int_{-\infty}^{\infty} e^{ixy - ty^2} \varphi(y) \, dy, \qquad \varphi(y) \in \{2, \sigma\},$ (1)

2. $c > \sigma$

$\Rightarrow \qquad |u(x, t)| \leq \dfrac{M(c)}{\sqrt{t - c}} e^{-x^2/4(t+c)}, \qquad c < t < \infty.$ (2)

Here $M(c)$ is independent of x and t but may depend on c.

Set $y = \xi + i\eta$ in a complex y-plane. By the hypothesis on $\varphi(y)$, there exists $N(c)$ independent of ξ and η, such that

$$|\varphi(\xi + i\eta)| \leq N(c) e^{c(\xi^2 + \eta^2)}.$$

9. A FURTHER CRITERION

Now shift the path of integration in (1) from the real axis, $\eta = 0$, to the line $\eta = A$. By Cauchy's theorem, this is permissible, for $t > c$ since

$$\left| \int_0^A \exp\left[ix(R + i\eta) - t(R + i\eta)^2 \right] \varphi(R + i\eta) \, d\eta \right|$$

$$\leqslant N(c) \int_0^A \exp(-x\eta + t\eta^2 - tR^2 + cR^2 + c\eta^2) |d\eta|,$$

and the right-hand side $\to 0$ as $|R| \to \infty$. For the new path of integration

$$u(x, t) = \frac{1}{2\pi} \int_{-\infty}^{\infty} \exp\left[ix(\xi + iA) - t(\xi + iA)^2 \right] \varphi(\xi + iA) \, d\xi,$$

$$|u(x, t)| \leqslant \frac{N(c)}{2\pi} \frac{[\exp(-Ax + A^2 t + A^2 c)]\sqrt{\pi}}{\sqrt{t - c}}. \tag{3}$$

Here we have replaced the probability integral by $\sqrt{\pi}$. The right-hand side of (3) is smallest when $A = x/[2(c + t)]$. Its value is then the right-hand side of (2) except for the factor $\sqrt{4\pi}$. If we set $M(c) = N(c)/\sqrt{4\pi}$, (2) is proved.

Corollary 9.1

$$u(x, t) \in H^o, \qquad \sigma < t < \infty.$$

Since differentiation under the integral sign (1) is obviously valid for $t > \sigma$, $u(x, t) \in H$. By Theorem 9.1 $u(x, t)$ is uniformly bounded in the half plane $t > c + \delta$, $\delta > 0$. Hence by Theorem 1, Chapter 7, it has a Poisson integral representation there, and by Corollary 1b, Chapter IX, belongs to H^Δ and hence to H^o there.

The converse of this corollary is false, as the example $u(x, t) \equiv 1$ shows. It cannot equal the Fourier integral (1) which $\to 0$ as $x \to 0$ by the Fourier–Lebesgue theorem, though it is a member of H^o.

Theorem 9.2

$\Leftrightarrow$ 1. $u(x, t) = \sum_{n=0}^{\infty} b_n w_n(x, t), \qquad 0 \leqslant \sigma < t,$ (4)

     A. $u(x, t) \in H^o, \qquad\qquad\qquad\qquad \sigma < t < \infty,$

     B. $\int_{-\infty}^{\infty} e^{x^2/(8t)} |u(x, t)| \, dx < \infty, \qquad \sigma < t < \infty.$

Assume first that (4) holds. Then by Theorem 8 equation (1) holds and A follows from Corollary 9.1. By Theorem 9.1,

$$\int_{-\infty}^{\infty} e^{x^2/(8t)} |u(x, t)| \, dx \leqslant \frac{M(c)}{\sqrt{t-c}} \int_{-\infty}^{\infty} e^{x^2/(8t) - x^2/4(t+c)} \, dx.$$

The dominant integral converges when $t > c$, so that B also follows. Conversely, condition A implies

$$u(x, t) = \int_{-\infty}^{\infty} k(x - y, t - s) u(y, s) \, dy, \qquad \sigma < s < t < \infty, \quad (5)$$

and 3(1) gives

$$k(x - y, t - s) = \sum_{n=0}^{\infty} \frac{v_n(y, -s) w_n(x, t)}{2^n n!}.$$

Substitution of this series in integral (5) clearly gives (4) with

$$b_n = \frac{1}{2^n n!} \int_{-\infty}^{\infty} u(y, s) v_n(y, -s) \, dy,$$

provided that term-by-term integration is valid. It is so if

$$\sum_{n=1}^{\infty} \frac{|w_n(x, t)|}{2^n n!} \int_{-\infty}^{\infty} |u(y, s) v_n(y, -s)| \, dy < \infty. \qquad (6)$$

By Theorem 2.2 and its corollary, series (6) is dominated by

$$\frac{e^{-x^2/(8t)}}{\sqrt{4\pi t}} \sum_{n=1}^{\infty} \left(\frac{e}{nt} \frac{es}{n} \right)^{n/2} n! \int_{-\infty}^{\infty} |u(y, s)| e^{y^2/(8s)} \, dy.$$

The test ratio of the series is $\sqrt{s/t}$ so that it converges, and the integral converges by condition B. Hence equation (4) is proved.

Corollary 9.2

$$1. \quad u(x, t) = \sum_{n=0}^{\infty} b_n w_n(x, t), \qquad t > \sigma,$$

$$\Rightarrow \qquad b_n = \frac{1}{2^n n!} \int_{-\infty}^{\infty} u(y, s) v_n(y, -s) \, dy, \qquad s > \sigma.$$

10 EXAMPLES

This integral appears to depend on s. That it does not do so follows from Corollary 8.

10 EXAMPLES

We illustrate the various series expansions of this chapter by a number of examples.

Example A

$$u(x, t) = e^{tD^2} e^{ax^2} = \frac{e^{ax^2/(1-4at)}}{\sqrt{1-4at}}.$$

This is the function of Lemma 2.1. Since it is positive, it belongs to H^o wherever it is defined: for $t < 1/(4a)$ when $a > 0$, for $t > 1/(4a)$ when $a < 0$, for all t when $a = 0$. Applying Theorem 6, we see that $u(x, t)$ must have an expansion in the $v_n(x, t)$ valid in the strip $|t| < 1/(4|a|)$ when $a \neq 0$. It is

$$u(x, t) = \sum_{n=0}^{\infty} \frac{a^n}{n!} v_{2n}(x, t).$$

If $a = 0$, the series reduces to the single term $v_0(x, t)$.

Applying Theorem 8, we observe first that $u(x, t)$ has the integral representation 8(2) if $a < 0$; for by Theorem 5, Chapter III, we have for $t > 1/(4a)$

$$u(x, t) = \frac{\sqrt{\pi}}{\sqrt{-a}} k\left(x, t - \frac{1}{4a}\right) = \frac{1}{\sqrt{-4\pi a}} \int_{-\infty}^{\infty} e^{ixy - ty^2 + y^2/(4a)} \, dy.$$

Hence,

$$\varphi(y) = \sqrt{\frac{\pi}{-a}} e^{y^2/(4a)} = \sqrt{\frac{\pi}{-a}} \sum_{n=0}^{\infty} \frac{y^{2n}}{(4a)^n n!}$$

$$u(x, t) = \sqrt{\frac{\pi}{-a}} \sum_{n=0}^{\infty} \frac{w_{2n}(x, t)}{4^{2n}(-a)^n n!}. \tag{1}$$

By inspection, $\varphi(y) \in \{2, 1/(4|a|)\}$, so that (1) is valid for $t > -1/(4a)$. We can check this by setting $x = 0$, when (1) becomes, except for unessential factors,

$$\sum_{n=0}^{\infty} \frac{(2n)!}{n! \, n!} \frac{1}{(16at)^n}.$$

It has test ratio $1/(4|a|t)$, as predicted.

Finally, the example illustrates Theorem 9.2. Condition B holds for $t > -1/(4a)$, for then

$$\int_{-\infty}^{\infty} \exp\left(\frac{x^2}{8t} + \frac{ax^2}{1-4at}\right) dx < \infty.$$

Example B The function $k(x - y, t + s)$ may be used to illustrate Theorems 6, 8, and 9.2. First consider y and s as parameters and set

$$u(x, t) = k(x - y, t + s).$$

Since $u \geq 0$, $u \in H^o$ for $t > -s$. By Theorem 6, there must be a polynomial expansion for $|t| < s$. By equation 1(4),

$$k(x - y, s) = \sum_{n=0}^{\infty} \frac{x^n}{2^n n!} w_n(y, s), \qquad 0 < s,$$

and

$$e^{tD^2} k(x - y, s) = k(x - y, s + t) = \sum_{n=0}^{\infty} \frac{v_n(x, t) w_n(y, s)}{2^n n!}, \qquad |t| < s. \tag{2}$$

The first equality holds since $k(x, t) \in H^o$, the second by Theorem 6. The result is a verification of Theorem 3, but not an independent one since (2) was used in the proof of Theorem 6.

Next take x and t as parameters and set

$$u(y, s) = k(x - y, t + s).$$

By Theorem 5, Chapter III,

$$u(y, s) = \frac{1}{2\pi} \int_{-\infty}^{\infty} \exp[i(x - y)z - (t + s)z^2] \, dz, \qquad t + s > 0,$$

so that the function φ of Theorem 8 is

$$\varphi(z) = e^{-ixz - tz^2}.$$

By inspection $\varphi(z) \in \{2, |t|\}$, so that there must be an expansion in terms of the $w_n(y, s)$ valid for $s > |t|$. By 1(3), $-iz$ replacing z,

$$\varphi(z) = \sum_{n=0}^{\infty} \frac{(-iz)^n}{n!} v_n(x, t). \tag{3}$$

By Corollary 8, the coefficients of the desired w_n-expansion are obtained from those of series (3) by dividing by $(-2i)^n$. Thus

$$u(y, s) = \sum_{n=0}^{\infty} \frac{w_n(y, s)v_n(x, t)}{2^n n!}, \qquad s > |t|.$$

This is a new verification of Theorem 3.

In illustration of Theorem 9.2, we note that

$$\int_{-\infty}^{\infty} k(x - y, s + t)e^{y^2/(8s)} \, dy < \infty$$

for $s > |t|$ since then

$$\frac{-y^2}{4(s + t)} + \frac{y^2}{8s} < 0.$$

Example C We may also use Blackman's function $B(x, t) = k(x, t + i)$ to illustrate Theorems 6, 8, and 9.2. By Theorem 2, Chapter IX, it belongs to H^o in the strip $-a < t < 1/a$ for any $a > 0$. The maximum strip of the form $|t| < \sigma$, in which a series of $v_n(x, t)$ converges, is obtained for $k(x, t + i)$ when $a = 1 = \sigma$. The coefficients of the series are obtained as follows:

$$B(x, 0) = \frac{e^{-x^2/(4i)}}{\sqrt{4\pi i}} = \frac{1}{\sqrt{4\pi i}} \sum_{n=0}^{\infty} \frac{i^n}{4^n} \frac{x^{2n}}{n!},$$

$$B(x, t) = e^{tD^2} B(x, 0) = \frac{1}{\sqrt{4\pi i}} \sum_{n=0}^{\infty} \frac{i^n}{4^n} \frac{v_{2n}(x, t)}{n!}, \qquad |t| < 1.$$

It is interesting to check the width of the strip of convergence at $x = 0$. Since $v_{2n}(0, t) = t^n (2n)!/n!$, we seek the radius of convergence of the power series

$$\frac{1}{\sqrt{4\pi i}} \sum_{n=0}^{\infty} \frac{(2n)!(it)^n}{4^n n! \, n!}.$$

It is clearly 1 by the ratio test. Indeed the series is the binomial expansion of $[4\pi(t + i)]^{-1/2}$, as expected.

To use the same example as illustration for Theorem 8, we see by Theorem 5, Chapter III, that

$$k(x, t + i) = \frac{1}{2\pi} \int_{-\infty}^{\infty} e^{ixy - ty^2} e^{-iy^2} \, dy,$$

so that

$$\varphi(y) = e^{-iy^2} \in \{2, 1\}.$$

By the theorem the w_n-expansion must converge for $t > 1$. Since

$$\varphi(y) = \sum_{n=0}^{\infty} \frac{(-i)^n}{n!} y^{2n},$$

Theorem 8 yields

$$k(x, t + i) = \sum_{n=0}^{\infty} \frac{w_{2n}(x, t)}{(-4i)^n n!}, \quad t > 1.$$

Again checking at $x = 0$, we have

$$w_{2n}(0, t) = \frac{1}{\sqrt{4\pi t}} \frac{(2n)!}{n!} \frac{1}{(-t)^n},$$

$$\frac{1}{\sqrt{4\pi(t + i)}} = \frac{1}{\sqrt{4\pi t}} \sum_{n=0}^{\infty} \frac{(2n)!}{n! \, n!} \left(-\frac{i}{4t}\right)^n.$$

This is an alternative binomial expansion valid for $|t| > 1$.

Finally, to apply Theorem 9.2, we must investigate when

$$\int_{-\infty}^{\infty} e^{x^2/(8t)} \left| e^{-x^2/4(t+i)} \right| dx < \infty.$$

This is clearly true when

$$\frac{1}{8t} - \frac{t}{4(t^2 + 1)} < 0$$

or when $t > 1$, as expected.

For a summary of some of the results of this chapter and of others see Haimo [1973].

Chapter XI

ANALOGIES

1 INTRODUCTION

There are many analogies between the theory of analytic functions and that of heat conduction. These may have appeared from time to time in earlier chapters. We shall try to make them more precise here, thus capitalizing on greater familiarity with the classical theory and perhaps producing a clearer perspective for the present theory.

The principal analogy which we wish to set up is between functions $u(x, t)$ belonging to H in a strip $a < t < b$ of the x, t-plane and functions $f(x)$ of the *real* variable x analytic on an interval $a < x < b$. Analyticity at a point x_0 means that $f(x)$ has a power series expansion valid in a neighborhood of that point. This series provides an analytic extension of $f(x)$ into the complex z-plane, $z = x + iy$, so that

$$f(z) = \sum_{n=0}^{\infty} \frac{f^{(n)}(x_0)(z - x_0)^n}{n!}$$

inside some disk $|z - x_0| < \rho$.

In this section we set down in Table I the objects to be compared, analytic functions on the left, temperature functions on the right. In the corresponding sections to follow we elucidate the items here briefly noted.

TABLE I

1. Singular function $$f(x) = 1/x$$	Source solution $$k(x, t) = \frac{e^{-x^2/(4t)}}{\sqrt{4\pi t}}$$
2. Inversion $$I[f(x)] = \frac{1}{x} f\left(\frac{1}{x}\right)$$	Appell transformation $$Ap[u(x, t)] = k(x, t) u\left(\frac{x}{t}, -\frac{1}{t}\right)$$
3. Monomials $$v_n(x) = x^n$$	Heat polynomials $$v_n(x, t) = n! \sum_{k=0}^{[n/2]} \frac{x^{n-2k}}{(n-2k)!} \frac{t^k}{k!}$$
4. Inverse monomials $$w_n(x) = I[x^n] = \frac{1}{x^{n+1}}$$	Associated functions $$w_n(x, t) = Ap[v_n(x, t)]$$ $$= k(x, t) v_n(x, -t) t^{-n}$$
5. Restricted analyticity, $f \in A^o$ $$f(x) = \frac{1}{2\pi i} \int_\Gamma \frac{f(z)}{x - z} dz,$$ Γ, a restricted circle of the complex z-plane	Huygens property, $u \in H^o$ $$u(x, t) =$$ $$\int_{-\infty}^{\infty} k(x-y, t-s) u(y, s) \, dy, \quad s < t$$
6. $e^{cD} f(x) = f(x + c)$ $f \in A^o$ in (a, b), $a < x < x + c < b$	$e^{cD^2} u(x, t) = u(x, t + c)$ $u \in H^o$ in (a, b), $a < t < t + c < b$
7. Biorthogonality $$\frac{1}{2\pi i} \int_\Gamma v_m(z) w_n(z) \, dz = \delta_{m,n}$$	Biorthogonality $$\frac{1}{n! \, 2^n} \int_{-\infty}^{\infty} v_m(x, -t) w_n(x, t) \, dx$$ $$= \delta_{m,n}$$
8. Generating functions $$\frac{1}{x - r} = \sum_{n=0}^{\infty} w_n(x) r^n, \quad |r| < |x|$$ $$\frac{1}{t - x} = \sum_{n=0}^{\infty} v_n(x) w_n(t), \quad |x| < |t|$$	Generating function $$k(x - r, t) = \sum_{n=0}^{\infty} \frac{w_n(x, t) r^n}{2^n n!}$$ $$k(x - y, t + s) = \sum_{n=0}^{\infty} \frac{v_n(x, t) w_n(y, s)}{2^n n!}$$
9. Maclaurin expansion $$f(x) = \sum_{n=0}^{\infty} a_n v_n(x)$$	Polynomial expansion $$u(x, t) = \sum_{n=0}^{\infty} a_n v_n(x, t)$$
10. Inverse expansion $$f(x) = \sum_{n=0}^{\infty} b_n w_n(x)$$	Associated expansion $$u(x, t) = \sum_{n=0}^{\infty} b_n w_n(x, t)$$
11. Criterion for polynomial expansion In the largest interval $|x| < \rho$ where $f \in A^o$	Criterion for polynomial expansion In widest strip $|t| < \sigma$ where $u \in H^o$
12. Criterion for inverse expansion A. Valid for $|x| > \rho$ if $f = I[g]$, $g \in A^o$ for $|x| < 1/\rho$ B. Valid for $x > \rho$ if $$f(x) = \int_0^{\infty} e^{-xr} \varphi(r) \, dr,$$ $$\varphi \in \{1, \rho\}$$	Criterion for associated expansion A. Valid for $t > \sigma$ if $u = Ap[g]$, $g \in H^o$ for $|t| < 1/\sigma$ B. Valid for $x > \sigma$ if $$u(x, t) = \int_{-\infty}^{\infty} e^{ixr - tr^2} \varphi(r) \, dr,$$ $$\varphi \in \{2, \sigma\}$$

3. HEAT POLYNOMIALS

We have seen repeatedly how basic is the role of the source solution in the theory of heat conduction, for example in the Poisson integral

$$u(x, t) = \int_{-\infty}^{\infty} k(x - y, t)u(y, 0)\, dy = k(x, t) * u(x, 0).$$

Thus, many temperature functions $u(x, t)$ may be represented in a half-plane $t > 0$ in terms of their values $u(x, 0)$ on the boundary of that region by means of a convolution of $u(x, 0)$ with the source solution. The singular function $1/z$ shares this property in the theory of analytic functions. Cauchy's classical formula (with clockwise integration)

$$f(x) = \frac{1}{2\pi i} \int_\Gamma \frac{f(z)}{x - z}\, dz = \frac{1}{z} * f(z),$$

where Γ is the unit circle, for example, gives the values of f on a diameter of that circle in terms of its values (after analytic continuation) on the circumference thereof, again by convolution. In our analogies it is consequently natural to set up a correspondence between $1/x$ and $k(x, t)$. Note the similar formulas:

$$\frac{1}{2\pi i} \int_\Gamma \frac{dz}{a - z} = 1, \quad \text{every } a, \quad -1 < a < 1;$$

$$\int_{-\infty}^{\infty} k(x - y, t)\, dy = 1, \quad \text{every } t, \quad 0 < t < \infty.$$

2 THE APPELL TRANSFORMATION

The inversion operator I,

$$I[f(x)] = f\left(\frac{1}{x}\right)\frac{1}{x},$$

frequently carries functions $f(x) \in A$, $|x| < \rho$, into functions which belong to A for $|x| > \rho$. This is certainly true if $f(z)$, the analytic continuation of $f(x)$ into the complex plane, $\in A$ in the disk $|z| < \rho$. The corresponding property for the Appell transformation is that if $u(x, t) \in H$ in the half-plane $-\infty < t < 0$, then $\text{Ap}[u(x, t)] \in H$ for $0 < t < \infty$. We proved this in Chapter I, §6.

3 HEAT POLYNOMIALS

The monomials $v_n(x) \in A$, $-\infty < x < \infty$. They correspond to the heat polynomials $v_n(x, t) \in H$, $-\infty < t < \infty$. There is also complete correspondence with respect to differentiation:

$$Dv_n(x) = nv_{n-1}(x), \qquad Dv_n(x, t) = nv_{n-1}(x, t).$$

Note also the parallel relations:

$$v_n(x) = \frac{1}{2\pi i} \int_\Gamma \frac{z^n}{x-z} \, dz = \frac{1}{z} * z^n,$$

$$v_n(x, t) = \int_{-\infty}^{\infty} k(x-y, t) y^n \, dy = k(x, t) * x^n.$$

4 ASSOCIATED FUNCTIONS

The two formulas for $w_n(x, t)$ given in the table are seen to be equivalent since $v_n(x, t^2)$ is homogeneous of degree n:

$$v_n(\lambda x, \lambda^2 t) = \lambda^n v_n(x, t).$$

Replacing t by $-t$ and λ by $1/t$, we obtain the second formula for $\mathrm{Ap}[v_n(x, t)]$. Again we have correspondence under differentiation:

$$D^n \frac{1}{x} = \frac{(-1)^n n!}{x^{n+1}}, \qquad D^n k(x, t) = \frac{w_n(x, t)}{(-2)^n};$$

$$D^n w_0(x) = (-1)^n n! \, w_n(x), \qquad D^n w_0(x, t) = \frac{w_n(x, t)}{(-2)^n}.$$

See 5(9), Chapter IX, for the last of these formulas.

5 THE HUYGENS PROPERTY

We have seen in Chapter IX how to restrict functions of H, in a useful way, so as to have the Huygens property and thus to belong to the smaller class H°. We now make precise a similar restriction, hinted at in item 5 of the table, for functions $f(x) \in A$.

Definition 5 $f(x) \in A^\circ$ in $a < x < b$ if it is the restriction to the real axis of a function which is analytic in a disk of the complex plane whose diameter is that interval.

An example will clarify this definition. The function $1/(x + i) \in A$ for $-\infty < x < \infty$ since it has a power series expansion about every point on the real axis. It belongs to A° on $-1 < x < 1$ since $1/(z + i) \in A$ for $|z| < 1$. It belongs to A° in no wider interval symmetric about the origin since there is a singularity in the complex plane at $z = -i$, and Taylor's theorem assures convergence of the series in powers of z out to the nearest singularity and no farther. The same theorem shows that $1/(x + i) \in A^o$ in $-\delta < x < 1/\delta$ for any $\delta > 0$ and in particular in $(0, \infty)$ or in $(-\infty, 0)$. This example shows that a function may belong to A^o in two overlapping intervals without doing so in their union. It is the analogue of Blackman's example (Chapter IX, §2) for temperature functions. By Cauchy's theorem functions $f(x)$ of class A^o have a representation by convolution with $1/z$ as follows:

$$f(x) \in A^o \quad \text{in } a < x < b \Rightarrow f(x) = \frac{1}{2\pi i} \int_\Gamma \frac{f(z)}{x - z} dz,$$

where Γ is any circle in the z-plane surrounding the point x and inside the circle whose diameter is the interval (a, b), and where integration is clockwise.

6 THE OPERATORS e^{cD} AND e^{cD^2}

We have already defined these two operators in Chapter III, §9. We have seen how easily the second is applied to functions of H^o:

$$e^{cD^2} u(x, t) = u(x, t + c),$$

provided only that (x, t) and $(x, t + c)$ lie inside the strip where $u \in H^o$. Indeed this follows essentially from the definition of the operator.

In a similar way

$$e^{cD} f(x) = f(x + c),$$

if only x and $x + c$ lie inside the interval (a, b) where $f \in A^o$. Then

$$f(x) = \frac{1}{2\pi i} \int_\Gamma \frac{f(z)}{x - z} dz, \quad f(x + c) = \frac{1}{2\pi i} \int_\Gamma \frac{f(z)}{x + c - z} dz,$$

where Γ must enclose both points x and $x + c$. Of course the second formula fails if $x + c$ is outside the interval where $f \in A^o$. For example, the integral has the value zero when $x + c = 2$ and $f(x) = 1/(x + i)$.

7 BIORTHOGONALITY

We proved the biorthogonality of the two sets of function $v_n(x, -t)$ and $w_n(x, t)$, $n = 0, 1, 2, \ldots$, in Chapter X, §1. The relationship was basic in the series expansions of functions of H. The corresponding formula for analytic functions is classic:

$$\frac{1}{2\pi i} \int_\Gamma v_m(z)w_n(z)\, dz = \frac{1}{2\pi i} \int_\Gamma \frac{z^m}{z^{n+1}}\, dz = \delta_{m,n},$$

where Γ is any circle enclosing the origin and integration is counterclockwise.

8 GENERATING FUNCTIONS

In Chapter X, §3, we proved the two formulas in the right-hand column of our table, the first valid for all x and r and for $0 < t$, the second for all x and y and for $0 \leq |t| < s$. The two formulas on the left are the same geometric series, written differently to emphasize the analogies.

9 POLYNOMIAL EXPANSIONS

Because the region of convergence of a power series is an interval $|x| < \rho$, we would expect from the parallelism which we have set up that the region of convergence of a series of heat polynomials would be a strip $|t| < \sigma$. We showed this to be the case in Chapter X, §4. There is strict analogy for the determination of the coefficients

$$f(x) = \sum_{n=0}^\infty a_n x^n, \qquad u(x,t) = \sum_{n=0}^\infty a_n v_n(x,t),$$
$$n!\, a_n = D^n f(x)\,|_{x=0}, \qquad n!\, a_n = D^n u(x,t)\,|_{(x,t)=(0,0)}.$$

There is close correspondence between the two formulas for determining ρ and σ:

$$\frac{1}{\rho} = \overline{\lim_{n\to\infty}} |a_n|^{1/n}, \qquad \frac{1}{\sigma} = \overline{\lim_{n\to\infty}} \left(\frac{2n}{e}\right)|a_n|^{2/n}.$$

The former is classical, the latter comes from Theorem 5.1, Chapter X.

10 ASSOCIATED FUNCTION EXPANSIONS

Corresponding to the region of convergence $|x| > \rho$ for the Laurent expansion

$$\sum_{n=0}^\infty b_n w_n(x),$$

that for a series of associated functions

$$\sum_{n=0}^{\infty} b_n w_n(x, t) \tag{1}$$

is the half-plane $t > \sigma$ (see Chapter X, §5). Again, ρ and σ are determined by analogous formulas:

$$\rho = \varlimsup_{n \to \infty} |b_n|^{1/n}, \qquad \sigma = \varlimsup_{n \to \infty} \left(\frac{2n}{e} \right) |b_n|^{2/n}.$$

We do not ordinarily consider series (1) for negative t, when the w_n are imaginary.

11 CRITERIA FOR POLYNOMIAL EXPANSIONS

As we have pointed out earlier, a temperature function $u(x, t)$ can be expanded in a series of heat polynomials in the widest strip $|t| < \sigma$ in which $u(x, t) \in H^o$, just as a function $f(x)$ can be expanded in powers of x in the widest interval $|x| < \rho$ in which $f(x) \in A^o$. In order to stress the analogy, let us state the two corresponding theorems.

Theorem 11.1

$$1. \quad f(x) = \sum_{n=0}^{\infty} a_n x^n, \qquad |x| < \rho,$$

$\Leftrightarrow$

$$f(x) \in A^o, \qquad |x| < \rho.$$

Theorem 11.2

$$1. \quad u(x, t) = \sum_{n=0}^{\infty} a_n v_n(x, t), \qquad |t| < \sigma,$$

$\Leftrightarrow$

$$u(x, t) \in H^o, \qquad |t| < \sigma.$$

12 CRITERIA FOR EXPANSIONS IN SERIES OF ASSOCIATED FUNCTIONS

There are two kinds of criteria for the expansion of an analytic function in negative powers of its variable. We state them both.

Theorem 12.1

$$1. \quad f(x) = \sum_{n=0}^{\infty} b_n w_n(x), \quad |x| > \rho,$$

$\Leftrightarrow \quad f(x) = I[g(x)] \quad$ some $g(x)$, $g(x) \in A^o$ in $|x| < 1/\rho$,

$$n!\, b_n = D^n g(x)|_{x=0}.$$

Theorem 12.2

$$1. \quad f(x) = \sum_{n=0}^{\infty} b_n w_n(x), \quad |x| > \rho, \tag{1}$$

$\Leftrightarrow \quad f(x) = \int_0^\infty e^{-xy} \varphi(y)\, dy \quad$ some $\varphi(y)$, $\varphi(y) \in \{1, \rho\}$,

$$b_n = D^n \varphi(y)|_{y=0}.$$

The proof of Theorem 21.1 is trivial and is thoroughly familiar. That of Theorem 12.1 is less so. We supply it here. Since

$$w_n(x) = \frac{1}{x^{n+1}} = \frac{1}{n!} \int_0^\infty e^{-xy} y^n \, dy, \quad x > 0,$$

we have formally

$$f(x) = \int_0^\infty e^{-xy} \varphi(y)\, dy, \quad \varphi(y) = \sum_{n=0}^{\infty} \frac{b_n}{n!} y^n.$$

The term-by-term integration can be checked as in the proof of Theorem 8, Chapter X. Series (1) converges for $|x| > \rho$ if and only if

$$\varlimsup_{n \to \infty} \sqrt[n]{|b_n|} \leq \rho. \tag{2}$$

From Boas [1954;11] an entire function

$$\sum_{n=0}^{\infty} a_n z^n$$

has growth $\{1, \sigma\}$ if and only if $\varlimsup_{n \to \infty} n|a_n|^{1/n} \leq \sigma e$. Hence from (2) we see that for $\varphi(y)$, with $a_n = 1/n!$,

$$\varlimsup \frac{n|b_n|^{1/n}}{e(n!)^{1/n}} = \varlimsup \frac{n|b_n|^{1/n}}{e[n^n e^{-n} \sqrt{2\pi n}]^{1/n}} \leq \rho.$$

12. CRITERIA FOR EXPANSIONS IN SERIES OF ASSOCIATED FUNCTIONS   203

That is, $\varphi(y) \in \{1, \rho\}$. Here we have used Stirling's formula. The theorem is proved.

The analogue of Theorem 12.1 is

Theorem 12.3

1. $$u(x, t) = \sum_{n=0}^{\infty} b_n w_n(x, t), \qquad t > \rho, \qquad (3)$$

$\Leftrightarrow \quad u(x, t) = \text{Ap}[v(x, t)], \qquad$ some $v(x, t), v(x, t) \in H^o$ in $|t| < 1/\rho$,

$$n! \, b_n = D^n v(x, t) \big|_{(x, t) = (0, 0)}.$$

This is an immediate consequence of Theorem 6, Chapter X, since $v(x, t) \in H^o$ for $|t| < 1/\rho$

$\Leftrightarrow \qquad\qquad v(x, t) = \sum_{n=0}^{\infty} b_n v_n(x, t), \qquad |t| < 1/\rho, \qquad (4)$

$$n! \, b_n = D^n v(x, t) \big|_{(x, t) = (0, 0)}.$$

Now apply the Appell transformation to both sides of (4), recalling that

$$\text{Ap}[v_n(x, t)] = w_n(x, t)$$

and that the strip $-1/\rho < t < 0$ goes into the half-plane $t > \rho$. Thus series (4) becomes series (3), and the proof is complete.

The analogue of Theorem 12.2 follows.

Theorem 12.4

1. $$u(x, t) = \sum_{n=0}^{\infty} b_n w_n(x, t), \qquad t > \rho,$$

$\Leftrightarrow \quad u(x, t) = \dfrac{1}{2\pi} \int_{-\infty}^{\infty} e^{ixy - ty^2} \varphi(y) \, dy, \qquad$ some $\varphi(y), \varphi(y) \in \{2, \rho\}$

$$b_n = \frac{D^n \varphi(y)}{n!(-2i)^n} \bigg|_{y=0}.$$

This is Theorem 8 and its corollary, Chapter X.

Chapter XII

HIGHER DIMENSIONS

1 INTRODUCTION

Thus far we have been dealing with the heat equation in a two-dimensional phase space of one space variable and one time variable. Physical problems in heat conduction usually lead to a differential equation involving three space variables. We have seen how this general equation may reduce to the simpler one already considered in detail when the physical setup is specialized so as to render the heat flow one-dimensional. Of course the type of possible problems increases enormously as the number of dimensions increases, due largely to the more involved topology. Many useful treatises (such as Carslaw and Jaeger [1948]) exist which specialize in such problems. It is not our purpose here to follow such a course. It is our intention rather to point out that most of our earlier results could be generalized, in a routine way, to higher dimensions, replacing intervals by squares or cubes and simple integrals by multiple integrals. In this chapter we shall derive the heat equation for solids and then illustrate how the generalization is done in the case of one important result, the criterion for polynomial expansions. The difficulties are largely notational, and to keep these to a minimum we treat the case of two space variables. The result is true in n dimensions; see Widder [1961b; 408].

The example chosen for treatment here is also of interest because it exhibits a phenomenon which could not have been observed in lower dimensions. We shall show that functions having the Huygens property H^o (generalized to the higher dimensions) can be approximated by linear combinations of polynomials each of which is factorable into two-dimensional heat polynomials. Thus these factorable polynomials may be thought of as generating a vector space or linear manifold whose closure, with respect to a suitable topology, includes H^o. Of course, members of H^o are not generally factorable into solutions of the two-dimensional heat equation.

2 THE HEAT EQUATION FOR SOLIDS

Let us begin by replacing the heat postulates of Chapter I by equivalent ones that will be more easily generalized to higher dimensional situations and which indeed will be more tractable mathematically.

If $u(x, t) \in C^2$ and is the temperature of a homogeneous insulated bar on the x-axis of an x, t-plane, then from 1(1), Chapter I, we have by the usual methods of the integral calculus that instantaneous rate of increase of the quantity $\mathcal{Q}$ in an arbitrary segment (a, b) of the bar is

$$\frac{\partial \mathcal{Q}}{\partial t} = c\rho \int_a^b \frac{\partial u}{\partial t} \, dx. \tag{1}$$

We may take this equation as a replacement for Postulate A. It describes the physical situation more clearly than the original postulate and, of course, reduces thereto when $\partial u / \partial t$ is constant over (a, b):

$$\frac{\partial \mathcal{Q}}{\partial t} = c\rho \frac{\partial u}{\partial t} (b - a) = cm \frac{\partial u}{\partial t}.$$

In like manner, we may substitute an integral expression for Postulate B. By 1(2), Chapter I, the rate at which heat is leaving the segment at b is $-lu_x(b, t)$ per unit area of cross section. The rate of entering at $x = a$ is $-lu_x(a, t)$, so that the rate of heat accumulation in the segment is

$$\frac{\partial \mathcal{Q}}{\partial t} = lu_x(b, t) - lu_x(a, t) = l \int_a^b u_{xx} \, dx. \tag{2}$$

Here we have used the fundamental theorem of the integral calculus, to be replaced by Green's theorem in higher dimensions. If equation (2) is taken as the replacement for Postulate B, then the heat equation for the bar is obtained immediately by equating the two values of $\partial \mathcal{Q}/\partial t$ from (1) and (2), as follows:

$$\int_a^b [lu_{xx}(x, t) - c\rho u_t(x, t)] \, dx = 0.$$

Since the integrand is continuous and since the segment (a, b) was arbitrary, the integrand must vanish identically, and we have recaptured equation 1(3), Chapter I.

Adopting the same physical model as in Chapter I, we may now phrase our postulates in an integral form suitable for any distribution of temperature in a solid. We assume that $u(x, y, z, t) \in C^2$ and is the temperature in a homogeneous body of constant density ρ, specific heat c, and thermal conductivity l. Here (x, y, z) are rectangular coordinates of a point inside the body and t is a time coordinate, as before.

Postulate A The instantaneous rate of accumulation of the quantity $\mathcal{Q}$ of heat in an arbitrary portion V of the above described body is

$$\frac{\partial \mathcal{Q}}{\partial t} = c\rho \iiint_V \frac{\partial u}{\partial t} \, dV. \tag{3}$$

Postulate B The instantaneous rate of dissemination of the quantity $\mathcal{Q}$ of heat from an arbitrary portion V of the above body through its surface Σ is

$$\frac{\partial \mathcal{Q}}{\partial t} = -l \iint_\Sigma \frac{\partial u}{\partial n} \, d\Sigma, \tag{4}$$

where $\partial u/\partial n$ is in the direction of the exterior normal to Σ.

These two postulates give in mathematical detail all of the facts intended by our description of the physical model given in Chapter I. They are in a form from which the derivation of the heat equation is almost immediate. If α, β, γ are the direction angles of the exterior normal to Σ, then

$$\frac{\partial u}{\partial n} = \frac{\partial u}{\partial x} \cos \alpha + \frac{\partial u}{\partial y} \cos \beta + \frac{\partial u}{\partial z} \cos \gamma.$$

By Green's theorem [Widder, 1961; 235], the surface integral (4) becomes the triple integral

$$\frac{\partial \mathcal{Q}}{\partial t} = -l \iiint_V \left[\frac{\partial^2 u}{\partial x^2} + \frac{\partial^2 u}{\partial y^2} + \frac{\partial^2 u}{\partial z^2} \right] dV. \tag{5}$$

The rate of accumulation of heat in V is the negative of (5), so that from (3) and (5) we have

$$\iiint_V \left[c\rho \frac{\partial u}{\partial t} - l\, \Delta u \right] dV = 0, \qquad \Delta u = \frac{\partial^2 u}{\partial x^2} + \frac{\partial^2 u}{\partial y^2} + \frac{\partial^2 u}{\partial z^2}.$$

Since the integrand is continuous and since V is arbitrary, we obtain the heat equation

$$\frac{\partial u}{\partial t} = \Delta u,$$

adopting the conventions of Chapter I about the physical constants.

To avoid complications that are notational only and for clarity of exposition we shall henceforth assume that u is independent of z. The equation then reduces to

$$\frac{\partial u}{\partial t} = \frac{\partial^2 u}{\partial x^2} + \frac{\partial^2 u}{\partial y^2}.$$

The results for this special case can easily be generalized to the general case, or indeed to n-dimensional space.

3 NOTATIONS AND DEFINITIONS

We consider a phase space with rectangular coordinates x, y, t. By the inequalities $a < t < b$ we mean the slab

$$\{(x, y, t) \mid -\infty < x < \infty, \quad -\infty < y < \infty, \quad a < t < b\}.$$

Definition 3.1

$$u(x, y, t) \in H, \qquad a < t < b,$$

$\Leftrightarrow$ A. $u(x, y, t) \in C^1, \qquad a < t < b,$

      B. $u_t = u_{xx} + u_{yy}, \qquad a < t < b.$

By B it is clear that u_{xx} plus u_{yy} exists and is continuous. A basic result is that we may obtain a function of H as the product of the two solutions of the one-dimensional heat equation, as follows.

Theorem 3.1

$$u(x, t),\ v(y, t) \in H, \qquad a < t < b,$$

$\Rightarrow$

$$u(x, t)v(y, t) \in H, \qquad a < t < b.$$

This is true since

$$(uv)_{xx} = vu_{xx}, \qquad (uv)_{yy} = uv_{yy},$$

$$(uv)_t = uv_t + vu_t = uv_{yy} + vu_{xx} = (uv)_{xx} + (uv)_{yy}.$$

We now define the *source solution* having source of unit strength at $(x, y) = (0, 0)$.

Definition 3.2

$$K(x, y, t) = k(x, t)k(y, t), \qquad -\infty < t < \infty.$$

We use the $*$ to indicate the two-dimensional convolution

$$u(x, y) * v(x, y) = \int_{-\infty}^{\infty} \int_{-\infty}^{\infty} u(x - \xi, y - \eta)v(\xi, \eta) \, d\xi \, d\eta.$$

Then the Poisson transform of the function $f(x, y)$ is defined as

$$u(x, y, t) = K(x, y, t) * f(x, y). \tag{1}$$

The addition formula for $K(x, y, t)$ becomes

$$K(x, y, t_1) * K(x, y, t_2) = K(x, y, t_1 + t_2). \tag{2}$$

By Definition 3.2 the double integral (2) becomes the product of two simple integrals to each of which Theorem 3, Chapter III, is applicable.

Definition 3.3

$$u(x, y, t) \in H^o \quad \text{(has the Huygens property)}, \quad a < t < b,$$

$$\Leftrightarrow \quad \begin{array}{ll} \text{A.} & u(x, y, t) \in H, \quad a < t < b, \\ \text{B.} & u(x, y, t) = K(x, y, t - t') * u(x, y, t'), \end{array} \tag{3}$$

the integral converging absolutely, for every t and t', $a < t' < t < b$.

For example, if $u(x, y, t)$ is defined by (1) with the integral converging absolutely in $a < t < b$, then $u \in H^o$ there, for then (3) follows from (2) and from Fubini's theorem in an obvious way.

We now define the heat polynomials.

Definition 3.4

$$v_{m,n}(x, y, t) = v_m(x, t)v_n(y, t), \qquad m, n = 0, 1, 2, \ldots \,.$$

4. GENERATING FUNCTIONS

Here v_n is the two-dimensional heat polynomial

$$v_n(x, t) = \int_{-\infty}^{\infty} k(x - y, t) y^n dy, \qquad t > 0. \tag{4}$$

Then

$$v_{m,n}(x, y, t) = K(x, y, t) * (x^m y^n), \qquad t > 0. \tag{5}$$

Again the double integral (5) becomes the product of two simple integrals to each of which (4) applies. Since (5) is a special case of (1) and converges absolutely, it is clear that $v_{m,n} \in H^o$, at least for $t > 0$. Indeed $v_{m,n} \in H^o$ for all t, as one sees by direct use of Definition 3.3 and the corresponding fact for $v_n(x, t)$.

Definition 3.5

$$w_{m,n}(x, y, t) = w_m(x, t) w_n(y, t).$$

Here $w_n(x, t)$ is the Appell transform of $v_n(x, t)$, as studied in Chapter X. Clearly

$$w_{m,n}(x, y, t) = K(x, y, t) v_{m,n}\left(\frac{x}{t}, \frac{y}{t}, -\frac{1}{t}\right).$$

It is called the function *associated* with $v_{m,n}(x, y, t)$.

We adopt here the more usual convention that a double series converges if and only if every simple series formed from all of its terms converges absolutely. Then a convergent series can be summed either by rows or by columns or by any other way involving all terms.

4 GENERATING FUNCTIONS

We set down here the generating functions for the heat polynomials and for their associated functions.

Theorem 4.1

$$-\infty < x < \infty, \qquad -\infty < y < \infty, \qquad -\infty < t < \infty$$

$$\Rightarrow \qquad e^{x\xi + y\eta + t(\xi^2 + \eta^2)} = \sum_{m=0}^{\infty} \sum_{n=0}^{\infty} v_{m,n}(x, y, t) \frac{\xi^m}{m!} \frac{\eta^n}{n!}.$$

The general term of this double series is a product of a function of m by a function of n. Hence the double series is really the product of two simple series to which equation 1(3), Chapter X, applies. The result is immediate.

Theorem 4.2

1. $-\infty < x < \infty, \quad -\infty < y < \infty, \quad 0 < t < \infty$

$$\Rightarrow \quad K(x - 2\xi, y - 2\eta, t) = \sum_{m=0}^{\infty} \sum_{n=0}^{\infty} w_{m,n}(x, y, t) \frac{\xi^m}{m!} \frac{\eta^n}{n!}.$$

This, too, is an immediate consequence of equation 1(4), Chapter X, after $w_{m,n}$ is replaced by its two factors.

Theorem 4.3

1. $-\infty < x < \infty, \quad -\infty < y < \infty, \quad 0 < s < \infty, \quad -s < t < s$

$$\Rightarrow \quad K(x - \xi, y - \eta, s + t) = \sum_{m=0}^{\infty} \sum_{n=0}^{\infty} \frac{v_{m,n}(x, y, t) w_{m,n}(\xi, \eta, s)}{2^m 2^n m!\, n!}.$$

Again the double series is the product of two simple series each of which has a known sum by Theorem 3, Chapter X. The product of the sums is $K(x - \xi, y - \eta, s + t)$, as stated.

5 EXPANSIONS IN SERIES OF POLYNOMIALS

We turn now to the principal result of this chapter, a criterion that a temperature function should be expansible in a series of heat polynomials. The conclusion is completely analogous to that of Theorem 6, Chapter X. A function $u(x, y, t)$ has such an expansion for $|t| < \sigma$ if and only if it belongs to H^σ there. We need a preliminary result.

Lemma 5

1. $0 < x_0, \quad 0 < y_0, \quad 0 < t_0,$ (1)

2. $\sum_{m=0}^{\infty} \sum_{n=0}^{\infty} a_{m,n} v_{m,n}(x_0, y_0, t_0) \quad$ converges

$$\Rightarrow \quad |a_{m,n}| \leq M \left(\frac{e}{2mt_0} \right)^{m/2} \left(\frac{e}{2nt_0} \right)^{n/2}, \quad m, n = 0, 1, \ldots, \quad (2)$$

for some constant M.

5. EXPANSIONS IN SERIES OF POLYNOMIALS

By the convergence of series (1), its general term is bounded by some constant N so that

$$|a_{m,n}| \leq \frac{N}{v_m(x_0, t_0) v_n(y_0, t_0)}.$$

By 4(1) and 4(2), Chapter X, and by Stirling's formula as there used, we obtain the desired conclusion at once.

Theorem 5

1. $u(x, y, t) \in H^o$, $\quad |t| < \sigma$,

$\Leftrightarrow \quad u(x, y, t) = \sum_{m=0}^{\infty} \sum_{n=0}^{\infty} a_{m,n} v_{m,n}(x, y, t), \quad |t| < \sigma,$ (3)

where

$$a_{m,n} = \frac{1}{m! \, n!} \frac{\partial^{m+n}}{\partial x^m \partial y^n} u(x, y, t) \Big|_{(x, y, t) = (0, 0, 0)} \tag{4}$$

$$= \frac{1}{m! \, n! \, 2^{m+n}} \int_{-\infty}^{\infty} \int_{-\infty}^{\infty} u(x, y, -t) w_{m,n}(x, y, t) \, dx \, dy, \quad 0 < t < \sigma. \tag{5}$$

First assume the representation (3), valid for $|t| < \sigma$. To prove that $u \in H^o$ there we must show that for any s, where $0 < s < c$,

$$u(x, y, t) = K(x, y, t + s) * u(x, y, -s), \quad |t| < s, \tag{6}$$

the integral converging absolutely. Substituting series (3) in the integral (6) and recalling that $v_{m,n}(x, y, t) \in H^o$ in $-\infty < t < \infty$, we obtain

$$\sum_{m=0}^{\infty} \sum_{n=0}^{\infty} a_{m,n} \int_{-\infty}^{\infty} \int_{-\infty}^{\infty} K(x - \xi, y - \eta, t + s) v_{m,n}(\xi, \eta, -s) \, d\xi \, d\eta$$

$$= \sum_{m=0}^{\infty} \sum_{n=0}^{\infty} a_{m,n} v_{m,n}(x, y, t) = u(x, y, t),$$

provided that term-by-term integration is valid. It will be so if

$$\sum_{m=0}^{\infty} \sum_{n=0}^{\infty} |a_{m,n}| \int_{-\infty}^{\infty} \int_{-\infty}^{\infty} K(x - \xi, y - \eta, t + s)$$

$$\times |v_{m,n}(\xi, \eta, -s)| \, d\xi \, d\eta < \infty. \tag{7}$$

By Definitions 3.2 and 3.4 the double integral (7) is the product of two simple integrals, the one depending on m, the other on n. Moreover, series (7) is dominated by another in which $a_{m,n}$ is replaced by the upper bound (2), itself factorable. Thus the dominant double series is the product of two simple series. Each is essentially series 6(5), Chapter X, after a_n of that series is replaced by $[e/(2nt')]^{n/2}$, as was done there. Hence we conclude that (7) is valid for $|t| < s$. That is, (6) holds and $u \in H^o$ for $|t| < \sigma$.

Conversely, assume $u \in H^o$ for $|t| < \sigma$. By Definition 3.3

$$u(x, y, t) = K(x, y, t + s) * u(x, y, -s) \tag{8}$$

for any pair of numbers t, s for which $-\sigma < -s < t \leq s < \sigma$, the integral converging absolutely. In particular, if $x = y = 0$ and $t = s$, we have

$$\int_{-\infty}^{\infty} \int_{-\infty}^{\infty} K(\xi, \eta, 2s)|u(\xi, \eta, -s)| \, d\xi \, d\eta < \infty. \tag{9}$$

By Theorem 4.3

$$K(x - \xi, y - \eta, t + s) = \sum_{m=0}^{\infty} \sum_{n=0}^{\infty} \frac{v_{m,n}(x, y, t) w_{m,n}(\xi, \eta, s)}{2^m m! \, 2^n n!}.$$

Substituting this series in the integral (8), we obtain

$$u(x, y, t) = \sum_{m=0}^{\infty} \sum_{n=0}^{\infty} a_{m,n} v_{m,n}(x, y, t),$$

where

$$a_{m,n} = \frac{1}{2^m m! \, 2^n n!} \int_{-\infty}^{\infty} \int_{-\infty}^{\infty} u(\xi, \eta, -s) w_{m,n}(\xi, \eta, s) \, d\xi \, d\eta,$$

provided that term-by-term integration is valid. It will be so if

$$\sum_{m=0}^{\infty} \sum_{n=0}^{\infty} \frac{|v_{m,n}(x, y, t)|}{2^m m! \, 2^n n!} \int_{-\infty}^{\infty} \int_{-\infty}^{\infty} |u(\xi, \eta, -s)| \, |w_{m,n}(\xi, \eta, s)| \, d\xi \, d\eta < \infty. \tag{10}$$

By Corollary 2.2, Chapter X,

$$|w_n(\xi, \eta, s)| \leq \frac{e^{-(\xi^2 + \eta^2)/(8s)}}{4\pi s} \left(\frac{2e}{ms}\right)^{m/2} m! \left(\frac{2e}{ns}\right)^{n/2} n!. \tag{11}$$

5. EXPANSIONS IN SERIES OF POLYNOMIALS

Hence the convergence of the integral (10) follows from that of

$$\int_{-\infty}^{\infty} \int_{-\infty}^{\infty} e^{-(\xi^2+\eta^2)/(8s)} |u(\xi, \eta, -s)| \, d\xi \, d\eta,$$

proved by (9). When the general term of series (10) is replaced by its upper bound from (11) it becomes factorable into a function of m by a function of n, so that the convergence of the series depends on that of the simple series

$$\sum_{n=0}^{\infty} \frac{|v_n(x, t)|}{2^n} \left(\frac{2e}{ns}\right)^{n/2}.$$

But this series has the same convergence property as series 6(7), Chapter X, which was found to converge for $|t| < s$. That is, equation (3) is established for $|t| < s$, and hence for $|t| < \sigma$, with the coefficients determined by (5). To derive the alternate determination (4), set $t = 0$ in (3),

$$u(x, y, 0) = \sum_{m=0}^{\infty} \sum_{n=0}^{\infty} a_{m,n} x^m y^n.$$

The constants $a_{m,n}$ are the coefficients of Maclaurin's series, as in (4). This concludes the proof.

It may be of interest to state the above result in the context of function space. Define the space E as the set of all functions $u(x, y, t)$ belonging to H on the slab $-\sigma < t < \sigma$. Since $u_1, u_2 \in H$ implies $a_1 u_1 + a_2 u_2 \in H$ for any complex constants a_1, a_2, it follows that E is a linear vector space over the field of complex constants. The space E^o of functions which belong to H^o on $|t| < \sigma$ is a linear vector subspace of E. Further the space S of all linear combinations of the heat polynomials $v_{m,n}(x, y, t)$ is again a linear vector subspace of E^o. If now we introduce into E the topology of pointwise convergence we may state as a consequence of Theorem 5 that every point of E^o is a limit point of points of S. That is, E^o is in the closure of S.

It is noteworthy that S is generated by the heat polynomials, all of which are factorable into the product of a solution of $u_{xx} = u_t$ by a solution of $u_{yy} = u_t$. Most members of S and of E^o do not have this property since generally the sum of two factorable functions is not factorable. For example, e^{x+y+2t} and xy are both factorable, but their sum is not. Of course each is a member of E^o, so that their sum is also. In the following section, we will give a less trivial example of a nonfactorable function of H^o.

6 AN EXAMPLE

We give here an illustration of Theorem 5. It is provided by the function

$$u(x, y, t) = (1 - 4t^2)^{-1/2} e^{(tx^2 + ty^2 + xy)/(1 - 4t^2)}.$$

It is clearly not factorable since for $t = 0$ it reduces to e^{xy}. It belongs to H for $|t| < \frac{1}{2}$ as one could show by direct differentiation. We can avoid this and at the same time show that $u \in H^o$ by showing that

$$u(x, y, t) = K(x, y, t) * e^{xy}. \tag{1}$$

It was pointed out in §3 that functions defined by Poisson integrals have the Huygens property in their regions of absolute convergence. Explicitly, the integral (1) is

$$\int_{-\infty}^{\infty} k(x - \xi, t) \, d\xi \int_{-\infty}^{\infty} k(y - \eta, t) e^{\xi \eta} \, d\eta. \tag{2}$$

The inner integral is

$$e^{tD^2} e^{\xi y} = e^{y\xi + t\xi^2}, \qquad D = \frac{\partial}{\partial y}.$$

Hence,

$$u(x, y, t) = \int_{-\infty}^{\infty} k(x - \xi, t) e^{y\xi + t\xi^2} \, d\xi$$

$$= e^{-y^2/(4t)} \int_{-\infty}^{\infty} k(x - \xi, t) \exp\left[t(\xi + (y/2t))^2\right] d\xi$$

$$= e^{-y^2/(4t)} \int_{-\infty}^{\infty} k\left(x - r + y(2t)^{-1}, t\right) e^{tr^2} \, dr, \qquad r = \xi + \frac{y}{2t}. \tag{3}$$

But in Lemma 2.1, Chapter X, we evaluated an integral of this type:

$$\int_{-\infty}^{\infty} k(x - y, t) e^{Ay^2} \, dy = \frac{e^{Ax^2/(1 - 4At)}}{\sqrt{1 - 4At}}, \qquad 0 < t < \frac{1}{4A}. \tag{4}$$

Hence to evaluate the integral (3) we have only to set $A = t$ in (4) and replace x by $x + y/(2t)$. Since

$$-\frac{y^2}{4t} + \frac{t(x + y/(2t))^2}{1 - 4t^2} = \frac{tx^2 + ty^2 + xy}{1 - 4t^2},$$

6. AN EXAMPLE

we see that the integral (2) is equal to $u(x, y, t)$ for $|t| < \frac{1}{2}$, as stated.

To obtain the expansion 5(2) of the present function we may use the formula 5(3) for the coefficients. Equivalently, from the Maclaurin series

$$e^{xy} = \sum_{n=0}^{\infty} \frac{(xy)^n}{n!},$$

we find that

$$u(x, y, t) = \sum_{n=0}^{\infty} \frac{v_{n,n}(x, y, t)}{n!}, \quad |t| < \tfrac{1}{2}.$$

The double series has reduced to the simple series of its diagonal terms. The special case $x = y$ produces the interesting expansion

$$\frac{e^{x^2/(1-2t)}}{\sqrt{1-4t^2}} = \sum_{n=0}^{\infty} \frac{v_n^2(x, t)}{n!}, \quad |t| < \tfrac{1}{2}.$$

This equation may be checked by setting $x = 0$, and then it reduces to the familiar binomial expansion

$$\frac{1}{\sqrt{1-4t^2}} = \sum_{n=0}^{\infty} \frac{(2n)!}{n!\,n!} t^{2n}, \quad |t| < \tfrac{1}{2},$$

by use of the formulas

$$v_{2n+1}(0, t) = 0, \qquad v_{2n}(0, t) = \frac{(2n)!\, t^n}{n!}.$$

Chapter XIII

HOMOGENEOUS TEMPERATURE FUNCTIONS

1 INTRODUCTION

We return now to the case of two variables x, t and examine what functions $u(x, t)$ satisfy the heat equation and also are homogeneous of degree n:

$$u(\lambda x, \lambda^2 t) = \lambda^n u(x, t), \quad \lambda > 0. \tag{1}$$

In Chapter III, §4, we found a two-parameter family of such functions in the special case $n = -1$. Here we extend the discussion to include all integral degrees, positive or negative. We proceed at once to the definition of four sets of functions which belong to H and which satisfy equation (1).

Definition 1.1 The functions

$$v_n(x, t) = \int_{-\infty}^{\infty} k(x - y, t) y^n \, dy, \quad t > 0, \quad n = 0, 1, \ldots,$$

$$V_n(x, t) = \int_{-\infty}^{\infty} e^{xy + ty^2} y^n \, dy, \quad t < 0, \quad n = 0, 1, \ldots,$$

1. INTRODUCTION

are of the first kind.

Definition 1.2 The functions

$$h_n(x, t) = \int_0^\infty k(x - y, t) y^n \, dy, \qquad t > 0, \quad n = 0, 1, \ldots,$$

$$H_n(x, t) = \int_0^\infty e^{xy + ty^2} y^n \, dy, \qquad t < 0, \quad n = 0, 1, \ldots,$$

are of the second kind.

It was already proved in Chapter X that $v_n(x, t)$ satisfies (1). We give an alternative proof here, one that applies equally well to $h_n(x, t)$.

$$v_n(\lambda x, \lambda^2 t) = \int_{-\infty}^\infty k(\lambda x - y, \lambda^2 t) y^n \, dy, \qquad t > 0,$$

$$= \lambda^{n+1} \int_{-\infty}^\infty k(\lambda x - \lambda z, \lambda^2 t) z^n \, dz, \qquad y = \lambda z.$$

Since $k(x, t)$ is homogeneous of degree -1, the proof is complete. No change in the calculation is needed for $h_n(x, t)$. Similarly,

$$V_n(\lambda x, \lambda^2 t) = \int_{-\infty}^\infty e^{\lambda xy + t\lambda^2 y^2} y^n \, dy, \qquad t < 0,$$

$$= \lambda^{-n-1} \int_{-\infty}^\infty e^{xz + tz^2} z^n \, dz, \qquad z = \lambda y.$$

Thus V_n and H_n are homogeneous of degree $-n - 1$.

Theorem 1.1

1. $n = 0, 1, 2, \ldots$

$\Rightarrow$ $v_n(x, t), h_n(x, t)$ are homogeneous of degree n, $t > 0$,
$V_n(x, t), H_n(x, t)$ are homogeneous of degree $-n - 1$, $t < 0$.

As usual, we use primes to indicate differentiation with respect to x. Then for $n = 1, 2, \ldots$

$$h_n'(x, t) = \int_0^\infty k'(x - y, t) y^n \, dy = n \int_0^\infty k(x - y, t) y^{n-1} \, dy.$$

Here we have integrated by parts. Also,

$$h_0(x, t) = \int_{-x}^{\infty} k(y, t)\, dy, \qquad h'_0(x, t) = k(x, t).$$

Derivatives of V_n and H_n are obtained trivially. We record the results.

Theorem 1.2

$$\begin{aligned}
& v'_0 = 0, && h'_0 = k; && & \\
& v'_n = nv_{n-1}, && h'_n = nh_{n-1}, && n = 1, 2, \ldots, & t > 0; \\
& V'_n = V_{n+1}, && H'_n = H_{n+1}, && n = 0, 1, 2, \ldots, & t < 0.
\end{aligned}$$

There is a useful duality among these four sets of functions established by the Appell transformation:

$$\mathrm{Ap}[u(x, t)] = k(x, t) u\left(\frac{x}{t}, \frac{-1}{t}\right), \qquad t > 0.$$

Since

$$\mathrm{Ap}[e^{ax + a^2 t}] = k(x - 2a, t),$$

we obtain at once the following result:

Theorem 1.3

$$\qquad 1. \quad n = 0, 1, 2, \ldots$$

$$\Rightarrow \qquad \mathrm{Ap}[V_n] = \frac{v_n}{2^{n+1}}, \qquad \mathrm{Ap}[H_n] = \frac{h_n}{2^{n+1}}.$$

2 THE TOTALITY OF HOMOGENEOUS TEMPERATURE FUNCTIONS

We now show that the functions of Definitions 1.1 and 1.2 belong to H^o in their regions of definition. This was already proved for $v_n(x, t)$ in Theorem 5.1, Chapter IX, and a similar proof applies to $h_n(x, t)$. Since the integrals defining H_n and V_n are Laplace integrals, differentiation under

2. THE TOTALITY OF HOMOGENEOUS TEMPERATURE FUNCTIONS

the integral sign is valid. Hence both $\in H$ for $t < 0$. Both $\in H^\circ$ also since

$$\int_{-\infty}^{\infty} k(x - z, t - \delta) V_n(z, \delta) \, dz$$

$$= \int_{-\infty}^{\infty} k(x - z, t - \delta) \, dz \int_{-\infty}^{\infty} e^{zy + \delta y^2} y^n \, dy$$

$$= \int_{-\infty}^{\infty} e^{\delta y^2} y^n \, dy \int_{-\infty}^{\infty} k(x - z, t - \delta) e^{zy} \, dz$$

$$= \int_{-\infty}^{\infty} e^{\delta y^2} e^{xy + ty^2 - \delta y^2} y^n \, dy$$

$$= V_n(x, t), \qquad \delta < t < 0.$$

Fubini's theorem is applicable by the absolute convergence of all integrals. A similar proof holds for H_n.

Theorem 2.1

$$\Rightarrow \quad \begin{array}{l} 1. \quad n = 0, 1, 2, \ldots \\ v_n(x, t), h_n(x, t) \in H^\circ, \quad t > 0, \\ V_n(x, t), H_n(x, t) \in H^\circ, \quad t < 0. \end{array}$$

Let us obtain alternate forms for the first two functions in each set. Obviously $v_0(x, t) = 1$ and $v_1(x, t) = x$. In Chapter III, §8, we introduced the notation

$$l(x, t) = \operatorname{erfc} \frac{x}{\sqrt{4t}} = \frac{2}{\sqrt{\pi}} \int_{x/\sqrt{4t}}^{\infty} e^{-y^2} \, dy, \qquad t > 0.$$

Using it, we have

$$h_0(x, t) = \int_{-x}^{\infty} k(y, t) \, dy = \frac{1}{\sqrt{\pi}} \int_{-x/\sqrt{4t}}^{\infty} e^{-y^2} \, dy = \frac{l(-x, t)}{2}.$$

From the equation

$$\int_0^\infty e^{-(x-y)^2/(4t)}(x-y)\,dy = -2te^{-x^2/(4t)}$$

we have at once that

$$h_1(x,t) = xh_0(x,t) + 2tk(x,t).$$

If now we use Theorem 1.3 and apply the inverse Appell transformation

$$\mathrm{Ap}^{-1}[u(x,t)] = 4\pi k(ix,-t)u\left(\frac{-x}{t}, \frac{-1}{t}\right)$$

to the above results, we obtain the desired dual formulas. We summarize them.

Theorem 2.2

$$\begin{aligned}
v_0 &= 1, & v_1 &= x; \\
h_0 &= \frac{l(-x,t)}{2}, & h_1 &= xh_0 + 2tk, & t &> 0; \\
V_0 &= 2\pi k(ix,-t), & V_1 &= -\frac{x}{2t}V_0, & t &< 0; \\
H_0 &= \pi k(ix,-t)l(-x,-t), & H_1 &= -\frac{xH_0 + 1}{2t}, & t &< 0.
\end{aligned}$$

We now obtain the totality of homogeneous functions of arbitrary integral degree, positive or negative.

Theorem 2.3

1. $u(x,t) \in H$, $\quad t > 0$,
2. $u(x,t)$ is homogeneous of degree n, $\quad n = 0, 1, 2, \ldots$,

$\Rightarrow$
$$u(x,t) = Av_n(x,t) + Bh_n(x,t), \tag{1}$$

where A and B are arbitrary constants.

Set $\lambda = 1/\sqrt{4t}$ in 1(1), and set $z = x/\sqrt{4t}$ and $f(z) = u(z,\tfrac{1}{4})$. As in Chapter III, §4,

$$u(x,t) = f(z)(4t)^{n/2}.$$

2. THE TOTALITY OF HOMOGENEOUS TEMPERATURE FUNCTIONS

From the heat equation we obtain

$$f''(z) + 2zf'(z) - 2nf(z) = 0. \tag{2}$$

By Theorem 1.1, this differential equation, known as Hermite's, has two known solutions

$$f_1(z) = v_n(z, \tfrac{1}{4}), \qquad f_2(z) = h_n(z, \tfrac{1}{4}).$$

They are linearly independent. For suppose a constant $C \neq 0$ exists such that

$$f_1(z) = Cf_2(z). \tag{3}$$

This produces a contradiction when n is odd, since it gives

$$f_1(0) = 0 = C \int_0^\infty k(y, t) y^n \, dy \neq 0.$$

If n is even, $\neq 0$, we may differentiate (3) to revert to the even case by Theorem 1.2. For $n = 0$, (3) is contradictory by Theorem 2.2. Every solution of (2) must be a linear combination of $f_1(z)$ and $f_2(z)$, so that equation (1) follows.

Theorem 2.4

1. $u(x, t) \in H$, $t < 0$,
2. $u(x, t)$ is homogeneous of degree $-n - 1$, $n = 0, 1, 2, \ldots$,

$\Rightarrow$ $u(x, t) = AV_n(x, t) + BH_n(x, t).$

This result follows by applying the inverse Appell transformation to equation (1). It carries a function of H of degree n, $t > 0$, into a function of H of degree $-n - 1$, $t < 0$.

Theorem 2.5

1. $u(x, t) \in H$,
2. $u(x, t)$ is homogeneous, integral degree

$\Rightarrow$ $u(x, t) \in H^o.$

This is a consequence of Theorems 2.1, 2.3, and 2.4. As a result, there are no null solutions of the heat equation homogeneous of integral degree. This result also gives us a simpler proof of Theorem 5.2, Chapter IX. Since

$w_n(ix, -t)$ is homogeneous for $t < 0$, it clearly belongs to H^o there, or $w_n(x, t) \in H^o$ for $t > 0$. Explicitly,

$$w_n(ix, -t) = \frac{V_n(x, t)(2i)^n}{n!},$$

in accord with Theorem 2.4.

3 RECURRENCE RELATIONS

All the homogeneous functions of §1 can be computed successively from those of Theorem 2.2 by use of the following recurrence formulas.

Theorem 3

1. $n = 1, 2, \ldots$

$\Rightarrow$
$$v_{n+1} = xv_n + 2ntv_{n-1}, \qquad t > 0, \qquad (1)$$

$$h_{n+1} = xh_n + 2nth_{n-1}, \qquad t > 0, \qquad (2)$$

$$2tV_{n+1} + xV_n + nV_{n-1} = 0, \qquad t < 0, \qquad (3)$$

$$2tH_{n+1} + xH_n + nH_{n-1} = 0, \qquad t < 0. \qquad (4)$$

Euler's equations for the homogeneous functions v_{n+1} and V_{n-1} are

$$xv'_{n+1} + 2tv''_{n+1} = (n + 1)v_{n+1},$$

$$xV'_{n-1} + 2tV''_{n-1} = -nV_{n-1}.$$

These derivatives can be computed by Theorem 1.2, thus giving (1) and (3). The same calculations produce equations (2) and (4).

This theorem enables us to compute easily the following values, of use later:

$$v_{2n}(0, t) = \frac{(2n)! \, t^n}{n!}, \qquad v_{2n+1}(0, t) = 0;$$

$$h_{2n}(0, t) = \frac{v_{2n}(0, t)}{2}, \qquad h_{2n+1}(0, t) = \frac{(4t)^{(2n+1)/2} n!}{\sqrt{4\pi}}; \qquad (5)$$

$$V_{2n}(0, t) = \frac{\sqrt{-\pi t}\,(-4t)^{-n}(2n)!}{n!}, \qquad V_{2n+1}(0, t) = 0. \qquad (6)$$

4 CONTINUED FRACTION DEVELOPMENTS

Let us introduce a new sequence of homogeneous polynomials $\omega_n(x, t)$, which satisfy a functional equation similar to 3(1).

Definition 4

$$\omega_{-1}(x, t) = 0, \qquad \omega_0(x, t) = 1$$

$$\omega_n(x, t) = x\omega_{n-1}(x, t) + 2nt\omega_{n-2}(x, t), \qquad n = 1, 2, \ldots . \qquad (1)$$

For example, $\omega_1(x, t) = x$, $\omega_2(x, t) = x^2 + 4t$, and $\omega_3(x, t) = x^3 + 10xt$. Except for ω_{-1}, ω_0 and ω_1 these polynomials do not satisfy the heat equation. They will appear in the convergents of a continued fraction to be discussed below.

Let us consider first a continued fraction whose successive convergents are $v_n(x, t)/h_n(x, t)$. It is easily seen to be

$$\cfrac{1}{h_0 + \cfrac{2tk}{x + \cfrac{2t}{x + \cfrac{4t}{x + \cfrac{6t}{x + \cdots}}}}}. \qquad (2)$$

After the first two irregular numerators, 1 and $2tk$, the successive ones form the arithmetic progression $2t, 4t, 6t, \ldots$. The first convergent is $1/h_0 = v_0/h_0$, as desired; the second is $x/(xh_0 + 2tk) = v_1/h_1$ by Theorem 2.2. From these two convergents, one obtains the next in classical fashion by use of the first partial quotient $2t/x$:

$$\frac{xv_1 + 2tv_0}{xh_1 + 2th_0} = \frac{v_2}{h_2}.$$

Here we have used 3(1) and 3(2). Now using these equations in an obvious induction, we see that v_n/h_n is the $(n + 1)$th convergent of the continued fraction (2), as stated. We now show that this general convergent tends to a limit as $n \to \infty$ and thus show that (2) converges.

Theorem 4.1

$\Rightarrow$

1. $x > 0, \qquad t > 0$

$$\lim_{n \to \infty} \frac{v_n(x, t)}{h_n(x, t)} = 1.$$

By Definitions 1.1 and 1.2, we need only show that

$$\mathcal{D}_n = \frac{v_n(x, t)}{h_n(x, t)} - 1 = \frac{\int_{-\infty}^{0} k(x - y, t) y^n \, dy}{\int_{0}^{\infty} k(x - y, t) y^n \, dy} \to 0, \quad n \to \infty.$$

This is equivalent to

$$\lim_{n \to \infty} \frac{\int_{0}^{\infty} e^{(-2xy - y^2)/(4t)} y^n \, dy}{\int_{0}^{\infty} e^{(2xy - y^2)/(4t)} y^n \, dy} = 0.$$

Since for $0 < y < \infty$

$$e^{-(xy)/(2t)} y \leqslant \frac{2t}{ex}, \quad e^{2xy/(4t)} \geqslant 1,$$

we have

$$|\mathcal{D}_n| \leqslant \frac{2t}{ex} \frac{\int_{0}^{\infty} e^{-y^2/(4t)} y^{n-1} \, dy}{\int_{0}^{\infty} e^{-y^2/(4t)} y^n \, dy} = \frac{1}{ex\sqrt{t}} \frac{\Gamma(n/2)}{\Gamma[(n+1)/2]}.$$

Here we have set $y = \sqrt{4tr}$ to identify these integrals as gamma functions. Now an application of Stirling's formula yields the desired result. We have thus proved the following result.

Theorem 4.2

$$\Rightarrow \qquad 1 = \cfrac{1}{h_0(x, t) + \cfrac{2tk(x, t)}{x + \cfrac{2t}{x + \cfrac{4t}{x + \cdots}}}}. \qquad (3)$$

1. $x > 0$, $t > 0$

The $(n + 1)$th convergent is $v_n(x, t)/h_n(x, t)$.

This result enables us to obtain a continued fraction expansion for $H_0(x, t)$, $t < 0$. Set z equal to a part of the fraction (2) as follows:

$$z = \cfrac{1}{x + \cfrac{2t}{x + \cfrac{4t}{x + \cdots}}},$$

4. CONTINUED FRACTION DEVELOPMENTS

so that equation (3) becomes

$$z = \frac{1 - h_0(x, t)}{2tk(x, t)} = \frac{1 - l(-x, t)/2}{2tk(x, t)}$$

$$= [2 - l(-x, t)]e^{x^2/(4t)}\sqrt{\pi/(4t)}, \quad t > 0.$$

But by Theorem 2.2,

$$H_0(-x, -t) = \pi k(ix, t)l(x, t) = e^{x^2/(4t)}\sqrt{\pi/(4t)}\, l(x, t),$$

so that $z = H_0(-x, -t)$. Here we have used

$$l(x, t) = 2 - l(-x, t), \quad t > 0,$$

which follows from the equation

$$\int_{-\infty}^{\infty} e^{-y^2}\, dy = \int_{x}^{\infty} e^{-y^2}\, dy + \int_{-x}^{\infty} e^{-y^2}\, dy.$$

Hence we can prove the following result.

Theorem 4.3

$$1. \quad x > 0, \quad t > 0$$

$$\Rightarrow \quad H_0(-x, -t) = \cfrac{1}{x + \cfrac{2t}{x + \cfrac{4t}{x + \cdots}}}. \tag{4}$$

The nth convergent is $\omega_{n-1}(x, t)/v_n(x, t)$.

It remains only to compute the convergents of (4). The first two are $1/x = \omega_0/v_1$ and $x/(x^2 + 2t) = \omega_1/v_2$. To obtain the third convergent from these two and the next partial quotient $(4t)/x$, we proceed in classic manner,

$$\frac{x\omega_1 + 4t\omega_0}{xv_2 + 4tv_1},$$

and this is ω_2/v_3 by 3(1) and 4(1). The induction thus begun is now easily completed.

The chief interest in this expansion lies in the fact that so complicated a transcendental function as $H_0(-x, -t)$ can be expressed so simply and can be approximated by the rational functions $\omega_n(x, t)/v_{n+1}(x, t)$.

5 DECOMPOSITION OF THE BASIC FUNCTIONS

We show now that all of the homogeneous functions of this chapter can be very simply expressed in terms of polynomials and the two transcendental functions e^x and erfc x (or $k(x, t)$ and $l(x, t)$). The functions $v_n(x, t)$ are themselves polynomials. For the $V_n(x, t)$ we have

Theorem 5.1

1. $n = 0, 1, 2, \ldots,$ $t < 0$

$$\Rightarrow \qquad V_n(x, t) = (-2t)^{-n} v_n(x, -t) V_0(x, t). \qquad (1)$$

From the homogeneity of $v_n(x, t)$ we have

$$v_n(x, t) = t^n v_n\left(\frac{x}{t}, \frac{1}{t}\right) = t^n v_n\left(\frac{x}{t}, \frac{1}{t}\right) v_0(x, t). \qquad (2)$$

But now the Appell transform of (1) is (2). Hence, by Theorem 2.2, $V_n(x, t)$ is a rational function multiplied by the single transcendental $k(ix, -t)$, as predicted.

Theorem 5.2

1. $n = 0, 1, 2, \ldots,$ $t > 0$

$$\Rightarrow \qquad h_n(x, t) = v_n(x, t) h_0(x, t) + 2t\omega_{n-1}(x, t) k(x, t). \qquad (3)$$

We prove this by induction. It is true for $n = 0, 1$ by Theorem 2.2 and Definition 4. Assume equations (3) and

$$h_{n-1}(x, t) = v_{n-1}(x, t) h_0(x, t) + 2t\omega_{n-2}(x, t) k(x, t), \qquad (4)$$

which is (3) with n replaced by $n - 1$. Now multiply equation (3) by x and (4) by $2nt$, and add. By Theorem 3.1 and Definition 4, the sum is equation (3) with n replaced by $n + 1$, and the induction is complete. Thus $h_n(x, t)$ is a linear combination of $k(x, t)$ and $l(-x, t)$, the combining functions being polynomials, as predicted.

Theorem 5.3

1. $n = 0, 1, 2, \ldots,$ $t < 0$

$$\Rightarrow \qquad (-2t)^n H_n(x, t) = v_n(x, -t) H_0(x, t) + \omega_{n-1}(x, -t). \qquad (5)$$

7. SERIES OF POLYNOMIALS

The Appell transformation applied to (5) yields (3), so that the result is immediate. In (5) the only transcendental function on the right is $H_0(x, t) = \pi k(ix, -t)l(-x, -t)$, so that our contention is substantiated.

6 SUMMARY

Let us arrange the functions of this chapter as shown in Table II, indicating by an arrow the manner in which the operators D and Ap carry one function into another.

Except in one case D decreases the degree by 1. The exception is $DV_0 = 0$. The derivative of h_0 is not indicated in the diagram but $Dh_0 = k$, so that here again the degree is diminished by 1. The Appell transformation changes negative degree to positive degree, $-n - 1$ to n.

TABLE II

	First kind		Second kind	Degree
	$D\begin{bmatrix} v_n \\ v_{n-1} \end{bmatrix}\begin{matrix}\\ n\end{matrix}$ ←		→ $\begin{matrix} h_n \\ h_{n-1} \end{matrix}\begin{bmatrix} D \\ n \end{bmatrix}$ ←	$\begin{matrix} n \\ n-1 \end{matrix}$
$t > 0$	⋮		⋮	
	$D\begin{bmatrix} v_1 \\ v_0 \end{bmatrix}$ ←	2^{n+1} Ap	$\begin{matrix} h_1 \\ h_0 \end{matrix}\Big] D$ ←	$\begin{matrix} 1 \\ 0 \end{matrix}$
		Ap	Ap	
	$D\begin{bmatrix} V_0 \\ V_1 \end{bmatrix}$		$\begin{matrix} H_0 \\ H_1 \end{matrix}\Big] D$ ←	$\begin{matrix} -1 \\ -2 \end{matrix}$
$t < 0$	⋮		⋮	
	$D\begin{bmatrix} V_{n-1} \\ V_n \end{bmatrix}$		$\begin{matrix} H_{n-1} \\ H_n \end{matrix}\Big] D$ ←	$\begin{matrix} -n \\ -n-1 \end{matrix}$

7 SERIES OF POLYNOMIALS

In the remainder of this chapter we discuss series expansions in terms of the four types of homogeneous functions introduced in §1. Various criteria for the expansibility of functions in such series exist. See Widder [1968,

1969]. Among these we choose only those analogous to that given in Theorem 8, Chapter X. Although we gave much attention there to polynomial expansions, we give here a further criterion in the genre envisaged.

Theorem 7

$$1. \quad u(x, t) = \sum_{n=0}^{\infty} a_n v_n(x, t), \qquad 0 < t < \sigma, \qquad (1)$$

$$\Leftrightarrow \quad u(x, t) = \int_{-\infty}^{\infty} k(x - y, t)\varphi(y)\, dy, \qquad \varphi \in \left\{2, \frac{1}{4\sigma}\right\}, \qquad (2)$$

$$a_n = \frac{\varphi^{(n)}(0)}{n!}.$$

Assume first the integral representation (2), where

$$\varphi(y) = \sum_{n=0}^{\infty} a_n y^n \qquad (3)$$

$$\varlimsup_{n \to \infty} n|a_n|^{2/n} \leqslant \frac{2e}{4\sigma}. \qquad (4)$$

Substituting series (3) in integral (2), we obtain (1) if term-by-term integration is valid. It will be so if

$$\int_{-\infty}^{\infty} k(x - y, t) \sum_{n=0}^{\infty} |a_n|\, |y|^n\, dy < \infty$$

or

$$\int_{-\infty}^{\infty} k(x - y, t)\varphi^*(y)\, dy < \infty, \qquad (5)$$

where

$$\varphi^*(y) = \sum_{n=0}^{\infty} |a_n| y^n, \qquad y > 0,$$

$$= \sum_{n=0}^{\infty} |a_n|(-y)^n, \qquad y < 0.$$

7. SERIES OF POLYNOMIALS

Each of the two latter entire functions has growth $\{2, 1/(4\sigma)\}$ by (4), so that

$$\varphi^*(y) = O(e^{y^2/(4\theta\sigma)}), \qquad |y| \to \infty,$$

for any θ, $0 < \theta < 1$. This guarantees the convergence of the integral (5) for $t < \theta\sigma$ and hence for $t < \sigma$.

Conversely, the convergence of series (1) at $(0, c)$, $0 < c < \sigma$, implies by 4(3) and 4(4), Chapter X, that

$$a_n = O\left[\left(\frac{e}{2nc}\right)^{n/2}\right], \qquad n \to \infty.$$

That is,

$$\varlimsup_{n\to\infty} n|a_n|^{2/n} \leqslant \frac{e}{2c}.$$

Since this holds for all $c < \sigma$, inequality (4) must also hold. Now define $\varphi(y)$ by series (3) and compute the integral (2) as before. In the presence of (4) the value of the integral is the series (1) and hence is $u(x, t)$. Since $\varphi(y) \in \{2, 1/(4\sigma)\}$ by (4), the proof is complete.

Observe that the argument is valid in the limiting case when $\sigma = \infty$, $\varphi \in \{2, 0\}$. A case in point is $\varphi(y) = e^{ay}$, when the function (1) becomes the generating function of the heat polynomials:

$$e^{ax+a^2t} = \sum_{n=0}^{\infty} \frac{a^n v_n(x, t)}{n!}, \qquad -\infty < t < \infty.$$

Note that the convergence of series (1) for $0 < t < \sigma$ implies its convergence for $-\sigma < t \leqslant 0$, so that $u(x, t)$ as defined by the Poisson integral (2) for $t > 0$ may have its definition extended, as a function of H^o, into the negative half-plane.

Corollary 7

1. $u(x, t) \in H^o$, $\qquad 0 \leqslant t < \sigma$,
2. $u(x, 0) \in \{2, 1/(4\sigma)\}$

$\Rightarrow \qquad u(x, t) \in H^o, \qquad -\sigma < t < \sigma.$

This corollary answers a question originally posed by P. Appell: When can a temperature function be extended backward in time? The answer:

Always, when the initial temperature is an entire function of order 2 and of finite type.

8 FIRST KIND, NEGATIVE DEGREE

As may be expected, expansions in series of $V_n(x, t)$ may be obtained from series 7(1) by use of the Appell transformation.

Theorem 8

1. $$u(x, t) = \sum_{n=0}^{\infty} b_n V_n(x, t), \qquad -\infty < t < -\sigma \leq 0, \qquad (1)$$

$$\Leftrightarrow \qquad u(x, t) = \int_{-\infty}^{\infty} e^{xy + ty^2} \varphi(y) \, dy, \qquad \varphi \in \{2, \sigma\}, \qquad (2)$$

$$b_n = \frac{\varphi^{(n)}(0)}{n!}.$$

Since

$$\text{Ap}(e^{xy + ty^2}) = k(x - 2y, t),$$

equation (2) holds if and only if

$$\text{Ap}[u(x, t)] = \tfrac{1}{2} \int_{-\infty}^{\infty} k(x - y, t) \varphi\left(\frac{y}{2}\right) dy, \qquad \varphi\left(\frac{y}{2}\right) \in \left\{2, \frac{\sigma}{4}\right\}. \qquad (3)$$

By Theorem 7, (3) is valid if and only if

$$\text{Ap}[u(x, t)] = \sum_{n=0}^{\infty} a_n v_n(x, t)$$

$$= \sum_{n=0}^{\infty} 2^{n+1} a_n \, \text{Ap}[V_n(x, t)], \qquad 0 < t < \frac{1}{\sigma}, \qquad (4)$$

$$a_n = \frac{\varphi^{(n)}(0)}{2^{n+1} n!}. \qquad (5)$$

Now apply the inverse Appell transformation to (4) to obtain

$$u(x, t) = \sum_{n=0}^{\infty} 2^{n+1} a_n V_n(x, t), \qquad -\infty < t < -\sigma. \qquad (6)$$

Setting $b_n = 2^{n+1}a_n$, equation (6) becomes (1) and equation (5) becomes $b_n = \varphi^{(n)}(0)/n!$, as stated in the theorem.

Again the limiting case $\sigma = 0$ is valid and is illustrated by $\varphi(y) = e^{ay}$:

$$V_0(x + a, t) = \sum_{n=0}^{\infty} \frac{a^n}{n!} V_n(x, t), \qquad -\infty < t < 0.$$

This equation is also evident by Maclaurin's series since $V_0(x, t)$ is the restriction to reals of an entire function and $D^n V_0(x, t) = V_n(x, t)$.

9 SECOND KIND, POSITIVE DEGREE

Corresponding to Theorem 7 for series of functions of the first kind we have here for those of the second kind the following result.

Theorem 9

1. $$u(x, t) = \sum_{n=0}^{\infty} a_n h_n(x, t), \qquad 0 < t < \sigma, \qquad (1)$$

$\Leftrightarrow \quad u(x, t) = \int_0^{\infty} k(x - y, t)\varphi(y)\, dy, \qquad \varphi \in \{2, 1/(4\sigma)\}, \qquad (2)$

$$a_n = \frac{\varphi^{(n)}(0)}{n!}.$$

Assume first the integral representation (2) where $\varphi(y)$ is defined by 7(3) and 7(4). Substituting the series in integral (2) and using Definition 1.2, we obtain series (1). The validity of the substitution for $0 < t < \sigma$ is proved precisely as in §7.

The proof of the converse is here somewhat simpler than for that of Theorem 7. From the convergence of (1) at $(0, c)$, $0 < c < \sigma$, we have from 3(5)

$$u(0, c) = \tfrac{1}{2} \sum_{n=0}^{\infty} \frac{(2n)!}{n!} a_{2n} c^n + \frac{1}{\sqrt{4\pi}} \sum_{n=0}^{\infty} n!\, a_{2n+1}(4c)^{n+(1/2)},$$

and hence that

$$a_{2n} = O\left(\frac{n!}{c^n(2n)!}\right), \quad a_{2n+1} = O\left(\frac{1}{(4c)^{n+(1/2)} n!}\right), \quad n \to \infty.$$

By Stirling's formula

$$a_n = O\left(\left(\frac{e}{2cn}\right)^{n/2}\right),$$

$$\varlimsup_{n\to\infty} n|a_n|^{2/n} \leqslant \frac{e}{2c}.$$

The rest of the proof follows as for Theorem 7.

10 SECOND KIND, NEGATIVE DEGREE

The dual of Theorem 9 with respect to the Appell transformation follows.

Theorem 10

$\Leftrightarrow$

1. $u(x, t) = \sum_{n=0}^{\infty} b_n H_n(x, t), \quad -\infty < t < -\sigma \leqslant 0,$ (1)

$$u(x, t) = \int_0^{\infty} e^{xy+ty^2} \varphi(y)\, dy, \quad \varphi \in \{2, \sigma\},$' (2)

$$b_n = \frac{\varphi^{(n)}(0)}{n!}.$$

The proof follows as in §8 by applying the Appell transformation to (1) and (2) and using Theorem 9. Again the limiting case $\sigma = 0$ is valid, as illustrated by the example $\varphi = e^{ay}$:

$$H_0(x + a, t) = \sum_{n=0}^{\infty} \frac{a^n H_n(x, t)}{n!}, \quad -\infty < t < 0.$$

11 EXAMPLES

We give here four examples in illustration of expansions in series of our four types of functions. In each case we choose $\varphi(y) = e^{cy^2}$, which clearly has growth $\{2, |c|\}$.

A For Theorem 7, we have by Lemma 2.1, Chapter X,

$$u(x, t) = k(x, t) * e^{cx^2} = \frac{e^{cx^2/(1-4ct)}}{\sqrt{1-4ct}}, \quad 1 - 4ct > 0, \quad (1)$$

$$u(x, t) = \sum_{n=0}^{\infty} \frac{v_{2n}(x, t) c^n}{n!}, \quad 0 < t < \frac{1}{4|c|}. \quad (2)$$

This series of polynomials also results from Theorem 6, Chapter X, as we saw in Example A, §10 of that chapter. Indeed, we proved there that (2) is valid in the wider strip $|t| < 1/(4|c|)$, in accord with Corollary 7. Series (2) may be checked at $x = 0$ when it becomes the familiar binomial expansion

$$\frac{1}{\sqrt{1-4ct}} = \sum_{n=0}^{\infty} \frac{(2n)!}{n!\, n!} (ct)^n, \quad |t| < \frac{1}{4|c|}.$$

B For Theorem 8,

$$u(x, t) = \int_{-\infty}^{\infty} e^{xy+ty^2} e^{cy^2}\, dy = V_0(x, t + c) \quad (3)$$

$$= \sqrt{\pi}\, (-t - c)^{-1/2} e^{-x^2/(4t+4c)}, \quad t + c < 0,$$

$$V_0(x, t + c) = \sum_{n=0}^{\infty} \frac{V_{2n}(x, t) c^n}{n!}, \quad -\infty < t < -|c|. \quad (4)$$

Observe that for $c < 0$, the integral (3) has a larger region of convergence than the series (4) and can then be used to extend the definition of $u(x, t)$ as a function of H^o beyond the region of convergence of the series (4). We may check series (4) by use of 3(6):

$$\frac{1}{\sqrt{-t-c}} = \frac{1}{\sqrt{-t}} \sum_{n=0}^{\infty} \binom{2n}{n} \left(\frac{-c}{4t}\right)^n.$$

This is a familiar binomial expansion valid for $-\infty < t < -|c|$.

Series (4) may also be regarded as a Maclaurin expansion since $V_0(x, t)$, with fixed x, is an analytic function of t for $-\infty < t < 0$, as one sees at once from its integral definition. In fact the integral (3), with $c = 0$, converges in the left half of a complex t-plane, so that $V_0(x, t) \in A^o$,

$-\infty < t < 0$, in the notation of Chapter XI, §5. Thus, by Maclaurin's theorem,

$$V_0(x, t + c) = \sum_{n=0}^{\infty} \frac{\partial^n}{\partial t^n} \frac{V_0(x, t)c^n}{n!}$$

$$= \sum_{n=0}^{\infty} \frac{V_0^{(2n)}(x, t)c^n}{n!}$$

$$= \sum_{n=0}^{\infty} \frac{V_{2n}(x, t)c^n}{n!}.$$

C For Theorem 9,

$$u(x, t) = \int_0^{\infty} k(x - y, t)e^{cy^2} \, dy$$

$$= k(x, t)H_0\left(\frac{x}{2t}, \frac{4ct - 1}{4t}\right)$$

$$= \sum_{n=0}^{\infty} \frac{h_{2n}(x, t)c^n}{n!}, \quad 0 < t < \frac{1}{4|c|}.$$

By use of Theorem 2.2, $u(x, t)$ may be expressed in terms of $k(x, t)$ and erfc x. It is

$$u(x, t) = \frac{e^{cx^2/(1-4ct)}}{\sqrt{4t(1 - 4ct)}} \text{ erfc } \frac{-x}{\sqrt{4t(1 - 4ct)}},$$

and the check for $x = 0$ is

$$\frac{1}{\sqrt{1 - 4ct}} = \sum_{n=0}^{\infty} \binom{2n}{n}(ct)^n, \quad |t| < 1/4|c|.$$

D Finally, for Theorem 10,

$$u(x, t) = \int_0^{\infty} e^{xy + ty^2 + cy^2} \, dy = H_0(x, t + c)$$

$$= \sum_{n=0}^{\infty} \frac{H_{2n}(x, t)c^n}{n!}, \quad -\infty < t < -|c|.$$

Again this result can be checked by Maclaurin's theorem.

Chapter **XIV**

MISCELLANEOUS TOPICS

1 POSITIVE TEMPERATURE FUNCTIONS

In Theorem 3, Chapter VIII, we showed that a solution $u(x, t)$ of the heat equation which is ≥ 0 for $t > 0$ must have the Poisson–Stieltjes representation

$$u(x, t) = \int_{-\infty}^{\infty} k(x - y, t) \, d\alpha(y), \qquad \alpha(y) \in \uparrow. \tag{1}$$

It is natural to ask if there is a corresponding integral representation for those solutions which are ≥ 0 for $t < 0$. The answer is affirmative. Indeed the two integral representations are dual to each other under the Appell transformation.

Theorem 1.1

1. $u(x, t) \in H, \geq 0, \qquad -\infty < t < 0,$

$\Leftrightarrow \quad u(x, t) = \int_{-\infty}^{\infty} e^{xy + ty^2} \, d\alpha(y), \qquad -\infty < t < 0, \quad \alpha(y) \in \uparrow. \tag{2}$

Equation (2) will hold if and only if its Appell transformation holds:

$$\text{Ap}[u(x, t)] = \int_{-\infty}^{\infty} k(x - 2y, t) \, d\alpha(y)$$

$$= \int_{-\infty}^{\infty} k(x - y, t) \, d\alpha\left(\frac{y}{2}\right), \qquad t > 0. \tag{3}$$

Comparing with (1), we see that (3) holds if and only if $\text{Ap}[u(x, t)] \geq 0$, $t > 0$, and hence if $u(x, t) \geq 0$, $t < 0$. This concludes the proof. A change of variable shows that if $u(x, t) \geq 0$ on any half-plane $t < c$, then $u(x, t)$ has the Laplace integral representation (2) there, and conversely.

This theorem enables us to solve a problem first considered by Appell [1892]: Under what circumstances must we conclude that a solution of the heat equation, without singularities, is constant? Is there an analogue of Liouville's theorem for functions analytic and bounded over the whole complex plane? From the physical point of view, we might be tempted to suppose that if heat has been diffusing along an insulated infinite bar (without sources or sinks) from the beginning of time, the temperature of the bar would end by being constant. That this is false is seen by the function $\exp(x + t)$, for example. Probably the trouble with intuition here is a difficulty with terms like "infinite" bar and the "beginning of time." Evidently heat may flow into the bar "from infinity." There is, however, an analogue of Liouville's theorem, as we now show.

Theorem 1.2

$$\Rightarrow \qquad \begin{aligned} 1. \quad & u(x, t) \in H \cdot B, \qquad t < 0, \\ & u(x, t) = \text{constant}, \qquad t < 0. \end{aligned}$$

By hypothesis, a constant M exists for which

$$-M \leq u(x, t) \leq M, \qquad t < 0.$$

Then we may apply Theorem 1.1 to the functions $M - u(x, t)$ and $M + u(x, t)$ to obtain

$$M - u(x, t) = \int_{-\infty}^{\infty} e^{xy + ty^2} \, d\alpha(y), \qquad \alpha(y) \in \uparrow, \tag{4}$$

$$M + u(x, t) = \int_{-\infty}^{\infty} e^{xy + ty^2} \, d\beta(y), \qquad \beta(y) \in \uparrow. \tag{5}$$

Denote by $U(y)$ the step-function with a single unit jump at $y = 0$, so that

$$1 = \int_{-\infty}^{\infty} e^{xy+ty^2} \, dU(y).$$

Then from (4) and (5)

$$\int_{-\infty}^{\infty} e^{xy+ty^2} \, d[MU(y) - \alpha(y)] = \int_{-\infty}^{\infty} e^{xy+ty^2} \, d[\beta(y) - MU(y)].$$

Assuming, as we may, that $\alpha(y)$ and $\beta(y)$ are normalized, we have by the uniqueness of bilateral Laplace transform representations that

$$MU(y) - \alpha(y) = \beta(y) - MU(y)$$

$$2MU(y) = \alpha(y) + \beta(y). \tag{6}$$

If $y_0 \neq 0$, $0 < \delta < |y_0|$, we have from (6) that

$$\alpha(y_0 + \delta) - \alpha(y_0) + \beta(y_0 + \delta) - \beta(y_0) = 0. \tag{7}$$

Since $\alpha(y)$ and $\beta(y)$ are both nondecreasing, equation (7) shows that they are both constant for $y > 0$ and for $y < 0$. That is,

$$2u(x, t) = \int_{-\infty}^{\infty} e^{xy+ty^2} \, d[\beta(y) - \alpha(y)]$$

$$= \beta(0+) - \beta(0-) - \alpha(0+) + \alpha(0-).$$

Thus, $u(x, t)$ is constant, as we wished to prove.

That the boundedness assumed in hypothesis 1 cannot be replaced by one-sided boundedness is evident from the example $e^t \cosh x$. However, Hirschman [1952; 487] has imposed a condition on the growth of $u(x, 0)$ as $|x| \to \infty$ which, together with positiveness, also produces the conclusion that $u(x, t)$ is constant.

2 POSITIVE DEFINITE FUNCTIONS

To treat certain subclasses of positive temperature functions we need to use the *positive definite* functions of Bochner [1959; 92]. He defined them in terms of certain quadratic forms and then characterized them as Fourier–Stieltjes integrals of bounded nondecreasing functions. Since it is only this integral representation that concerns us here, let us take his characterization as our definition.

Definition 2

$$f(x) \in \text{PD} \quad \text{(positive definite)}$$

$$\Leftrightarrow \quad f(x) = \int_{-\infty}^{\infty} e^{ixy}\, d\alpha(y), \qquad \alpha(y) \in \uparrow \cdot B.$$

We list a number of illustrations

Example A

$$f(x) = \cos ax, \quad -\infty < a < \infty.$$

Here $\alpha(y)$ is a step-function with jumps at $x = \pm a$. The function $f(x) = 1$ is a special case.

Example B

$$f(x) = k(x, a), \quad a > 0.$$

By Theorem 5, Chapter III,

$$\alpha(y) = \frac{1}{2\pi} \int_{-\infty}^{y} e^{-ar^2}\, dr,$$

$$k(x, a) = \frac{1}{2\pi} \int_{-\infty}^{\infty} e^{ixy - ay^2}\, dy.$$

Example C

$$f(x) = e^{-ax^2}, \quad a > 0.$$

By Theorem 7, Chapter III,

$$\alpha(y) = \int_{-\infty}^{y} k(r, a)\, dr,$$

$$e^{-ax^2} = \int_{-\infty}^{\infty} e^{ixy} k(y, a)\, dy.$$

3. POSITIVE TEMPERATURE FUNCTIONS, CONCLUDED

Example D

$$e^{-|x|} = \frac{1}{\pi} \int_{-\infty}^{\infty} \frac{e^{ixy}}{y^2 + 1} \, dy.$$

Example E

$$\frac{2}{x^2 + 1} = \int_{-\infty}^{\infty} e^{ixy} e^{-|y|} \, dy.$$

The last two examples are familiar Fourier transforms. See, for example, Widder [1961c; 435].

3 POSITIVE TEMPERATURE FUNCTIONS, CONCLUDED

We now restrict the nonnegative members of H, $t > 0$, by demanding further that they should belong to L on horizontal lines. We show that all such functions are Fourier transforms in accordance with the following theorem.

Theorem 3.1

1. $u(x, t) \in H, \geq 0, \quad t > 0,$
2. $\int_{-\infty}^{\infty} u(x, t) \, dx < \infty, \quad t > 0,$

$\Leftrightarrow \quad u(x, t) = \dfrac{1}{2\pi} \int_{-\infty}^{\infty} e^{ixy - ty^2} \varphi(y) \, dy, \quad \varphi(y) \in \text{PD}. \quad (1)$

By Definition 2, equation (1) will hold if and only if

$$u(x, t) = \frac{1}{2\pi} \int_{-\infty}^{\infty} e^{ixy - ty^2} \, dy \int_{-\infty}^{\infty} e^{iyr} \, d\alpha(r), \quad \alpha(r) \in \uparrow \cdot B. \quad (2)$$

By Fubini's theorem, valid since $\alpha(r) \in \uparrow \cdot B$, and by Theorem 5, Chapter III, equation (2) is equivalent to

$$u(x, t) = \int_{-\infty}^{\infty} k(x + r, t) \, d\alpha(r) = \int_{-\infty}^{\infty} k(x - y, t) \, d[-\alpha(-y)], \quad (3)$$

$$\int_{-\infty}^{\infty} u(x, t) \, dx = \alpha(\infty) - \alpha(-\infty) < \infty. \quad (4)$$

But by Theorem 3, Chapter VIII, relations (3) and (4) hold if and only if the present hypotheses 1 and 2 do. This completes the proof.

In illustration of the theorem we have

$$k(x + a, t) + k(x - a, t) = \frac{1}{\pi} \int_{-\infty}^{\infty} e^{ixy - ty^2} \cos ay \, dy,$$

or

$$k(x, t + a) = \frac{1}{2\pi} \int_{-\infty}^{\infty} e^{ixy - ty^2} e^{-ay^2} \, dy, \qquad a > 0,$$

corresponding, respectively, to Examples A and C, §2.

The dual result under the Appell transformation follows.

Theorem 3.2

1. $u(x, t) \in H, \geq 0, \qquad t < 0,$

2. $\int_{-\infty}^{\infty} u(x, t) e^{x^2/(4t)} \, dx < \infty, \qquad t < 0,$

$\Leftrightarrow \qquad u(x, t) = \int_{-\infty}^{\infty} k(ix - y, -t) \varphi(y) \, dy, \qquad \varphi(y) \in \text{PD}. \qquad (5)$

Since

$$\text{Ap}[k(ix - 2y, -t)] = \frac{e^{ixy - ty^2}}{4\pi},$$

equation (5) will hold if and only if

$$\text{Ap}[u(x, t)] = \frac{1}{2\pi} \int_{-\infty}^{\infty} e^{ixy - ty^2} \varphi(2y) \, dy, \qquad t > 0.$$

Clearly $\varphi(2y) \in \text{PD}$ when $\varphi(y) \in \text{PD}$, so we have only to apply Theorem 3.1 to $\text{Ap}[u(x, t)]$. For this function to belong to H and to be nonnegative for $t > 0$ is equivalent to the present hypothesis 1, and

$$\int_{-\infty}^{\infty} k(x, t) u\left(\frac{x}{t}, -\frac{1}{t}\right) dx < \infty, \qquad t > 0,$$

is equivalent to hypothesis 2. This concludes the proof.

The example $\varphi(y) = 1$, $u(x, t) = 1$ provides an illustration of the theorem. Hypotheses are trivially satisfied, and $\varphi \in \text{PD}$.

As a less trivial example take $\varphi(y) = e^{-ay^2}$, $a > 0$. By Lemma 2.1, Chapter X,

$$u(x, t) = \int_{-\infty}^{\infty} k(ix - y, -t) e^{-ay^2} \, dy = \frac{e^{ax^2/(1-4at)}}{\sqrt{1-4at}}.$$

Here $u(x, t) \in H, \geq 0$ for $-\infty < t < 1/(4a)$. Also condition 2 holds since

$$\frac{a}{1-4at} + \frac{1}{4t} < 0, \qquad t < \frac{1}{4a},$$

and $\varphi \in PD$ by Example C, §2.

4 A STATISTICAL PROBLEM

In statistics a function $f(x)$ is called a *frequency function* if

$$f(x) \geq 0, \qquad \int_{-\infty}^{\infty} f(x) \, dx = 1. \tag{1}$$

Some of these functions have a *Gaussian representation*,

$$f(x) = \int_{-\infty}^{\infty} k(x - y, a) \, d\alpha_a(y), \qquad \alpha_a(y) \in \uparrow \tag{2}$$

for some $a > 0$ and some $\alpha_a(y) \in \uparrow$. By (1) and Fubini's theorem

$$\int_{-\infty}^{\infty} f(x) \, dx = \alpha_a(\infty) - \alpha_a(-\infty) = 1.$$

As an example, consider the frequency function $e^{-x^2}/\sqrt{\pi}$. The addition formula for $k(x, t)$ gives

$$\frac{e^{-x^2}}{\sqrt{\pi}} = k(x, \tfrac{1}{4}) = \int_{-\infty}^{\infty} k(x - y, a) k(y, \tfrac{1}{4} - a) \, dy,$$

so that this function has the representation (2) with

$$\alpha_a(y) = \int_0^y k(r, \tfrac{1}{4} - a) \, dy$$

for any a, $0 < a < \frac{1}{4}$. Indeed, if $U(y)$ is the unit step-function of §1

$$\frac{e^{-x^2}}{\sqrt{\pi}} = \int_{-\infty}^{\infty} k(x - y, \tfrac{1}{4})\, dU(y),$$

so that (2) holds even for $a = \frac{1}{4}$. Does it hold for $a > \frac{1}{4}$? We show later that the answer is negative, so that for this example $a = \frac{1}{4}$ is the maximum value of a for which (2) is valid. We prove that this example is typical; for an arbitrary frequency function $f(x)$ there is a maximum value of a for which (2) holds. For background material, as related to statistics, see Pollard [1953]. See also Pollard and Widder [1969].

Lemma 4

1. $f(x) = \int_{-\infty}^{\infty} k(x - y, b)\, d\alpha(y), \qquad b > 0, \quad \alpha(y) \in \uparrow \cdot B,$

$\Rightarrow$
$$f(x) \in \left\{ 2, \frac{1}{4b} \right\}. \tag{3}$$

Here we have used the notation of Definition 7, Chapter X. Of course $f(x)$, as here defined is real, but the notation (3) means that $f(x)$ is the restriction to reals of an entire function of $z = x + iy$, which has growth $\{2, 1/(4b)\}$. From hypothesis 1,

$$f(z) = k(z, b) \int_{-\infty}^{\infty} e^{zr/(2b)} e^{-r^2/(4b)}\, d\alpha(r), \tag{4}$$

$$|f(x + iy)| \leq \frac{e^{(y^2 - x^2)/(4b)}}{\sqrt{4\pi b}} \int_{-\infty}^{\infty} e^{(2xr - r^2)/(4b)}\, d\alpha(r).$$

Since

$$2xr - r^2 \leq x^2, \qquad -\infty < x < \infty, \quad -\infty < r < \infty,$$

$$|f(z)| \leq \frac{e^{y^2/(4b)}}{\sqrt{4\pi b}} \int_{-\infty}^{\infty} d\alpha(r) \leq \frac{e^{|z|^2/(4b)}}{\sqrt{4\pi b}} \int_{-\infty}^{\infty} d\alpha(r). \tag{5}$$

Integral (4) is a bilateral Laplace integral convergent for all z, so that $f(z)$ is entire. Inequality (5) gives (3) at once.

We can now impose conditions on $f(x)$ to ensure that it shall have the Gaussian representation (2) and that there shall exist a finite maximum value a for which (2) holds.

4. A STATISTICAL PROBLEM

Theorem 4

1. $f(x)$ is a frequency function,
2. $f(z)$ is entire, order 2, type $\dfrac{1}{4b}$, $z = x + iy$,
3. $u(x,t) = \sum\limits_{n=0}^{\infty} \dfrac{f^{(n)}(0)}{n!} v_n(x, t),$ \hfill (6)
4. $u(x, t) \geq 0,$ \hfill $-b < t \leq 0,$

$\Rightarrow \qquad f(x) = \int_{-\infty}^{\infty} k(x - y, a)\, d\alpha_a(y), \qquad \alpha_a(y) \in \uparrow$ \hfill (7)

for every a, $0 < a \leq b$, and for no $a > b$.

Here, as usual, $v_n(x,t)$ is the heat polynomial $k(x, t) * x^n$. We show first that series (6) converges for $|t| < b$. Setting $a_n = f^{(n)}(0)/n!$, we must prove, by Theorem 5.1, that

$$\varlimsup_{n \to \infty} \frac{|a_n|^{2/n}(2n)}{e} = \frac{1}{b}.$$ \hfill (8)

But a standard formula [Boas, 1954; 11] for the type of $f(x)$ gives

$$\varlimsup_{n \to \infty} \frac{n|a_n|^{2/n}}{2e} = \frac{1}{4b}.$$

This is equivalent to (8). By Theorem 6, Chapter X, $u(x, t) \in H^o$ for $|t| < b$. Since $v_n(x, 0) = x^n$, Taylor's theorem gives $u(x, 0) = f(x)$. By the definition of H^o,

$$u(x, t) = \int_{-\infty}^{\infty} k(x - y, t + a)u(y, -a)\, dy, \qquad -b < -a < t < b,$$

$$u(x, 0) = f(x) = \int_{-\infty}^{\infty} k(x - y, a)u(y, -a)\, dy, \qquad 0 < a < b, \hfill (9)$$

so that (7) holds for $0 < a < b$ with

$$\alpha_a(y) = \int_{-\infty}^{y} u(r, -a)\, dr. \hfill (10)$$

Clearly $\alpha_a(y) \in \uparrow$ by hypothesis 4. If (7) also held for $a > b$ we should have $f(x) \in \{2, 1/(4a)\}$ by Lemma 4. By hypothesis 2 the order is 2 so that the type must be $\leq 1/(4a) < 1/(4b)$, contradicting hypothesis 2.

It remains only to show that the maximum b is attained, that (7) holds with $a = b$. By (9) and (10) and hypothesis 1, we have $0 \leq \alpha_a(y) \leq 1$ for $0 < a < b$, so that we may apply Helly's theorem, Chapter VIII, §1, to be assured that there exists a sequence $a_1 < a_2 < \cdots < b$ tending to b and a function $\alpha_b(y)$ such that

$$\lim_{n \to \infty} \alpha_{a_n}(y) = \alpha_b(y), \quad -\infty < y < \infty.$$

Integrating (9) by parts, with a replaced by a_n, we have

$$f(x) = \int_{-\infty}^{\infty} k'(x - y, a_n) \alpha_{a_n}(y) \, dy$$

$$= \int_{-\infty}^{\infty} k'(x - y, b) \alpha_b(y) \, dy. \tag{11}$$

Here we have allowed n to become infinite and used Lebesgue's limit theorem, applicable since

$$|k'(x, a)| \leq \frac{2}{\sqrt{\pi}} \frac{|x|}{b^{3/2}} e^{-x^2/(4b)}, \quad \frac{b}{4} \leq a \leq b.$$

A final integration of (11) by parts shows that (7) holds for $a = b$, as stated.

5 EXAMPLES

According to Theorem 4, the growth of the variable a to larger values may be stopped by two possible factors, (a) the width of the strip $-b < t < 0$ into which $u(x, t)$ may be extended, backward in time, as a temperature function, and (b) a change of sign of $u(x, t)$ somewhere in the strip. We give here two examples to show that the two factors are independent.

Example A This is the frequency function $f(x) = k(x, \tfrac{1}{4})$ introduced in §4. Clearly $f(x)$ is entire, of order 2 and of type 1. Hence by Lemma 2.1, Chapter X,

$$u(x, t) = \frac{e^{-x^2/(1+4t)}}{\sqrt{\pi(1 + 4t)}}, \quad t > -\tfrac{1}{4}.$$

5. EXAMPLES

Obviously $u(x, t) \in H, \geq 0$ for $t > -\frac{1}{4}$. Since $u(x, t)$ is singular at $(0, -\frac{1}{4})$, $u(x, t) \notin H$ in any half-plane including that point. Hence it is factor (a) above that is operative. Theorem 4 assures that $b = \frac{1}{4}$ is the optimal choice of a, as seemed likely from preliminary considerations.

Example B

$$f(x) = ce^{-x^2/8}(x^2 + 4), \quad 1 = \int_{-\infty}^{\infty} f(x)\, dx.$$

The constant c has been chosen to make $f(x)$ a frequency function. To compute $u(x, t) = k(x, t) * f(x)$, we begin with the equation

$$k(x, t) * e^{-Ax^2} = \frac{e^{-Ax^2/(1+4At)}}{\sqrt{1 + 4At}}$$

from Lemma 2.1, Chapter X, differentiate with respect to A, and set $A = \frac{1}{4}$,

$$k(x, t) * (x^2 e^{-x^2/4}) = \frac{e^{-x^2/[4(t+1)]}(x^2 + 2t^2 + 2t)}{(t + 1)^{5/2}}.$$

Then

$$u(x, t) = c2^{5/2} k(x, t + 1) * (x^2 e^{-x^2/4})$$

$$= \frac{c2^{5/2} e^{-x^2/[4(t+2)]}(x^2 + 2(t + 1)(t + 2))}{(t + 2)^{5/2}},$$

$$u(x, 0) = f(x). \tag{1}$$

Observe that $u(x, t)$ can be extended backward in time as a function of H to $t = -2$, as might have been predicted from the fact that $f(x)$ is of order 2 and type $\frac{1}{8}$. So equation 4(9) holds for all a, $0 < a < 2$. But $u(x, t)$ is not ≥ 0 throughout that strip. For, the term $2(t + 1)(t + 2)$ appearing in (1) is > 0 for $-1 < t < 0$ but is < 0 for $-2 < t < -1$. Hence $u(x, t)$ is certainly negative near the t-axis in the strip $-2 < t < -1$. Hence the optimal choice for a in this case is $a = 1$, and it is factor (b) above that is operative:

$$ce^{-x^2/8}(x^2 + 4) = \int_{-\infty}^{\infty} k(x - y, 1) u(y, -1)\, dy.$$

Notice the marked contrast between these two examples. In Example A, the maximizing function $\alpha_{-1/4}(y)$ is the discontinuous function $U(y)$, whereas in Example B, it is the entire function

$$\alpha_{-1}(y) = c2^{5/2} \int_{-\infty}^{y} e^{-r^2/4} r^2 \, dr.$$

Theorem 4 is not applicable to this example for, if we try to apply it with $b = 1$, then hypothesis 4 is satisfied but 2 is not (the type is $\frac{1}{8}$, not $\frac{1}{4}$); if we try $b = 2$, when hypothesis 2 is satisfied, then 4 is not. In the following section we prove a more general theorem which will include both examples and which will always apply when a maximum value of a exists.

6 STATISTICAL PROBLEM CONCLUDED

For the definitive solution of our problem we use the positive definite functions of §2.

Lemma 6

$$\Rightarrow \quad \begin{array}{l} 1. \ \ f(x) \in \text{PD} \\[4pt] e^{-rx^2} f(x) \in \text{PD}, \quad r > 0. \end{array} \quad (1)$$

By hypothesis 1 and Example C, §2,

$$f(x) = \int_{-\infty}^{\infty} e^{ixy} \, d\alpha(y), \qquad \alpha(y) \in \uparrow \cdot B,$$

$$e^{-rx^2} = \int_{-\infty}^{\infty} e^{ixy} k(y, r) \, dy.$$

Set

$$\varphi(x) = \int_{-\infty}^{\infty} k(x - y, r) \, d\alpha(y),$$

so that $\varphi(x) \geq 0$ and $\in L(-\infty, \infty)$:

$$\int_{-\infty}^{\infty} \varphi(x) \, dx = \int_{-\infty}^{\infty} d\alpha(y) < \infty.$$

6. STATISTICAL PROBLEM CONCLUDED

Then

$$e^{-rx^2}f(x) = \int_{-\infty}^{\infty} e^{ixy}k(y, r)\, dy \int_{-\infty}^{\infty} e^{ixz}\, d\alpha(z)$$

$$= \int_{-\infty}^{\infty} d\alpha(z) \int_{-\infty}^{\infty} e^{ixy}k(y - z, r)\, dy$$

$$= \int_{-\infty}^{\infty} e^{ixy}\, dy \int_{-\infty}^{\infty} k(y - z, r)\, d\alpha(z)$$

$$= \int_{-\infty}^{\infty} e^{ixy}\varphi(y)\, dy = \int_{-\infty}^{\infty} e^{ixy}\, d\beta(y), \qquad \beta(x) = \int_{-\infty}^{x} \varphi(y)\, dy. \quad (2)$$

Since the product (1) has the representation (2) with $\beta(x) \in \uparrow \cdot B$, it is positive definite as stated. The above interchange of order of integration is valid since

$$\int_{-\infty}^{\infty} |d\alpha(z)| \int_{-\infty}^{\infty} k(y - z, r)\, dy = \int_{-\infty}^{\infty} d\alpha(z) < \infty.$$

We now introduce a subclass of frequency functions. We use the symbol $\hat{f}$ for the Fourier transform of f:

$$\hat{f}(x) = \int_{-\infty}^{\infty} e^{ixy}f(y)\, dy.$$

Definition 6

$$f(x) \in P_a$$

⇔ 1. $f(x)$ is a continuous frequency function,
   2. $\hat{f}(x)e^{rx^2} \in PD$ for $r < a$, $\notin PD$ for $r > a$.

The function of Example A, §5, belongs to $P_{1/4}$ because

$$\hat{f}(x) = \frac{1}{\sqrt{\pi}} \int_{-\infty}^{\infty} e^{ixy - y^2}\, dy = e^{-x^2}/4,$$

$$\hat{f}(x)e^{rx^2} = e^{-x^2/4}e^{rx^2} \in PD, \qquad r \leq \tfrac{1}{4},$$

$$\notin PD, \qquad r > \tfrac{1}{4}.$$

As a second example, $f(x) = e^{-|x|}/2 \in P_0$, for

$$\hat{f}(x)e^{rx^2} = \frac{e^{rx^2}}{x^2+1} \in \text{PD}, \quad r \leq 0,$$

$$\notin \text{PD}, \quad r > 0.$$

We show now that every continuous frequency function belongs to some class P_a, $0 \leq a < \infty$.

Theorem 6.1

1. $f(x)$ is a continuous frequency function

$\Rightarrow \quad f(x) \in P_a \quad$ some a, $\quad 0 \leq a < \infty$.

Observe first that if $s < r$, then

$$e^{rx^2}\hat{f}(x) \in \text{PD} \Rightarrow e^{sx^2}\hat{f}(x) \in \text{PD} \qquad (3)$$

for

$$\hat{f}(x)e^{sx^2} = e^{(s-r)x^2}\left(e^{rx^2}\hat{f}(x)\right),$$

so that (3) follows from Lemma 6. If the right-hand side of (3) is false, so is the left. Hence we may use the Dedekind cut, dividing all numbers r into two classes A and B according to whether the left-hand side of (3) is true or false, respectively. There are finite nonnegative numbers in each class. Obviously, $r = 0$ is in class A since $\hat{f} \in \text{PD}$:

$$\hat{f}(x) = \int_{-\infty}^{\infty} e^{ixy}f(x)\,dx, \quad f \geq 0, \quad f \in L.$$

Very large numbers r are in class B for, if for all $r > r_0$

$$\hat{f}(x)e^{rx^2} = \int_{-\infty}^{\infty} e^{ixy}\,d\alpha_r(y),$$

we should have in particular at $x = 0$, by 4(1),

$$1 = \int_{-\infty}^{\infty} d\alpha_r(y).$$

Hence

$$|\hat{f}(x)| \leq e^{-rx^2}$$

for every $r > r_0$, so that $f(x) \equiv 0$, contradicting 4(1). The existence of a finite point a dividing the two classes is thus established.

Theorem 6.2

1. $f(x) \in P_b$, $\quad 0 < b < \infty$,

$$\Leftrightarrow f(x) = \int_{-\infty}^{\infty} k(x - y, a) \, d\alpha_a(y), \quad \alpha_a(y) \in \uparrow, \quad \int_{-\infty}^{\infty} d\alpha_a(y) = 1, \quad (4)$$

for every a, $0 < a \leqslant b$, and for no a, $a > b$.

Under Fourier transformation, the convolution (4) becomes the product

$$\hat{f}(x) = e^{-ax^2} \int_{-\infty}^{\infty} e^{ixy} \, d\alpha_a(y) \quad (5)$$

by virtue of Example C, §2. (Proof is also immediate by use of Fubini's theorem.) That is, equation (4) is valid for a given value of a if and only if equation (5) holds. Comparison of (5) with Definition (6) now concludes the proof.

This theorem applied to Example A, §5, again confirms that the optimal value of a is $\frac{1}{4}$, since we saw above that this function is a member of $P_{1/4}$. For Example B,

$$f(x) = \int_{-\infty}^{\infty} k(x - y, t) u(y, -t) \, dy, \quad 0 < t < 2,$$

since $u(x, t) \in H^o$ for $t > -2$. Hence by the product theorem

$$\hat{f}(x) = e^{-tx^2} \int_{-\infty}^{\infty} e^{ixy} u(y, -t) \, dy.$$

Since $u(y, -t) \geqslant 0$ for $0 < t \leqslant 1$ and $\not\geqslant 0$ for $1 < t < 2$, it follows that $e^{tx^2}\hat{f}(x) \in$ PD for $0 < t \leqslant 1$ and $\notin$ PD for $1 < t < 2$. That is, $f \in P_1$ and the optimal value of a is 1, as we discovered directly in §5.

7 ALTERNATE INVERSION OF THE h-TRANSFORM

In Chapter IV, §7, we inverted the h-transform

$$u(x, t) = \int_0^t h(x, t - y) \varphi(y) \, dy.$$

The inversion required a knowledge of $u(x, t)$ in a neighborhood of the t-axis. We now inquire if inversion is still possible when $u(x, t)$ is known only on a single vertical line remote from the t-axis. For example, we seek to find $\varphi(y)$ if

$$f(t) = \int_0^t h(t - y)\varphi(y)\,dy, \qquad h(y) = h(1, y). \tag{1}$$

Blackman and Pollard [1959] restricted the problem further by assuming $f(t)$ known only in a neighborhood of $t = 0$. If on the contrary $f(t)$ is known for all $t > 0$ and if its Laplace transform $L[f(t)]$ exists, the problem has a familiar solution. For, by Theorem 8.1, Chapter III, $L[h] = e^{-\sqrt{s}}$, so that the product theorem for Laplace transforms gives

$$L[\varphi] = e^{\sqrt{s}} L[f],$$

and φ is then found as an inverse Laplace transform, from tables or otherwise. We give here an alternate solution of the restricted problem. See Pollard and Widder [1970].

Let us give first a heuristic derivation of our solution using the operational methods of Chapter III, §9. Let us use $\mathcal{D}$ as the symbol for differentiation with respect to t. From Theorem 8.1, Chapter III,

$$e^{-\sqrt{\mathcal{D}}}\varphi(t) = \int_0^\infty e^{-y\mathcal{D}}\varphi(t)h(y)\,dy$$

$$= \int_0^\infty \varphi(t - y)h(y)\,dy, \tag{2}$$

and assuming $\varphi(t) \equiv 0$ for $t < 0$, we have from equations (1) and (2) that $e^{-\sqrt{\mathcal{D}}}\varphi(t) = f(t)$. Thus the inversion operator should be $e^{\sqrt{\mathcal{D}}}$ in some sense. The following seems a reasonable definition for the operator.

Definition 7

$$e^{r\sqrt{\mathcal{D}}} f(t) = (\cosh r\sqrt{\mathcal{D}})f(t) + (\sinh r\sqrt{\mathcal{D}})f(t)$$

$$\cosh r\sqrt{\mathcal{D}} f(t) = \sum_{n=0}^\infty \frac{r^{2n}}{(2n)!} f^{(n)}(t)$$

$$\sinh r\sqrt{\mathcal{D}} f(t) = \sum_{n=1}^\infty \frac{r^{2n-1}}{(2n-1)!} g^{(n)}(t)$$

$$g(t) = f^{(-1/2)}(t) = \frac{1}{\sqrt{\pi}} \int_0^t \frac{f(y)}{\sqrt{t-y}}\,dy. \tag{3}$$

7. ALTERNATE INVERSION OF THE h-TRANSFORM

The definition of $g(t)$ is the classic Riemann–Liouville fractional integral of order $\frac{1}{2}$. The desired operator $e^{\sqrt{\mathcal{D}}}$ will appear by allowing r to approach 1 in the above. We begin by applying the operator to the kernel $h(t)$ of the transform (1).

Theorem 7.1

1. $h(1, t) = (4\pi)^{-1/2} t^{-3/2} e^{-1/(4t)}$,
2. $-\infty < r < \infty, \quad 0 < t < \infty$

$\Rightarrow \quad e^{r\sqrt{\mathcal{D}}} h(1, t) = h(1 - r, t).$

We compute (3) by the product theorem for Laplace transforms. Since $L[t^{-1/2}] = \sqrt{\pi/s}$, we have

$$L[g] = \frac{e^{-\sqrt{s}}}{\sqrt{s}}.$$

Hence by Theorem 7.2, Chapter III,

$$g(t) = \mathcal{D}^{-1/2} h(1, t) = 2k(1, t). \tag{4}$$

Continuing our convention that superscripts indicate differentiation with respect to the first variable we have

$$g^{(n)}(t) = 2k^{(2n)}(1, t) = -h^{(2n-1)}(1, t). \tag{5}$$

Hence

$$e^{r\sqrt{\mathcal{D}}} h(1, t) = \sum_{n=0}^{\infty} \frac{r^{(2n)}}{(2n)!} h^{(2n)}(1, t) - \sum_{n=0}^{\infty} \frac{r^{(2n-1)}}{(2n-1)!} h^{(2n-1)}(1, t). \tag{6}$$

Here we have used the fact that $h(x, t)$ satisfies the heat equation so that $(\partial/\partial t)^n h(1, t) = h^{(2n)}(1, t)$. The sum of series (6) is $h(1 - r, t)$ by Maclaurin's theorem, valid since $h(x, t)$ is an entire function of x for fixed $t \neq 0$.

Theorem 7.2

1. $\varphi(y) \in L, \quad 0 \leq y \leq c,$
2. $f(t) = \int_0^t h(1, t - y) \varphi(y) \, dy$

$\Rightarrow \quad \lim_{r \to 1-} e^{r\sqrt{\mathcal{D}}} f(t) = \varphi(t), \quad \text{almost all } t \text{ in } (0, c).$

If the inversion operator can be applied under the integral sign, we have by Theorem 7.1

$$e^{r\sqrt{\mathfrak{D}}} f(t) = \int_0^t h(1 - r, t - y)\varphi(y)\, dy. \tag{7}$$

To prove this we again use Maclaurin's theorem. We have by Theorem 9.1, Chapter IV, that the function

$$F(s) = \int_0^t h(s, t - y)\varphi(y)\, dy, \qquad s = \sigma + i\tau,$$

is analytic for $|\arg s| < \pi/4$. Thus

$$F(1 - r) = \sum_{n=0}^{\infty} \frac{F^{(n)}(1)(-r)^n}{n!} \tag{8}$$

for real r near $r = 1$ (actually for $|r - 1| < 1/\sqrt{2}$). Using classical rules for differentiating integrals and observing that $h^{(n)}(x, 0+) = 0$, we have

$$F^{(2n)}(1) = \int_0^t h^{(2n)}(1, t - y)\varphi(y)\, dy$$

$$= \int_0^t \left(\frac{\partial}{\partial t}\right)^n h(1, t - y)\varphi(y)\, dy = f^{(n)}(t).$$

Also

$$\mathfrak{D}^{-1/2} f(t) = \frac{1}{\sqrt{\pi}} \int_0^t \frac{dy}{\sqrt{t - y}} \int_0^y h(1, y - z)\varphi(z)\, dz$$

$$= \frac{1}{\sqrt{\pi}} \int_0^t \varphi(z)\, dz \int_z^t \frac{h(1, y - z)}{\sqrt{t - y}}\, dy$$

$$= 2 \int_0^t k(1, t - z)\varphi(z)\, dz.$$

Here we have used (4) and Fubini's theorem, applicable by hypothesis 1 and the fact that $h(1, t) > 0$ for $t > 0$. By (5)

$$g^{(n)}(t) = 2 \int_0^t k^{(2n)}(1, t - z)\varphi(z)\, dz$$

$$= -\int_0^t h^{(2n-1)}(1, t - z)\varphi(z)\, dz$$

$$= -F^{(2n-1)}(1).$$

Hence equation (8) is equivalent to

$$F(1 - r) = \sum_{n=0}^{\infty} f^{(n)}(t) \frac{r^{2n}}{(2n)!} - \sum_{n=1}^{\infty} g^{(n)}(t) \frac{r^{2n-1}}{(2n - 1)!}$$

$$= e^{r\sqrt{\mathcal{D}}} f(t).$$

Equation (7) is thus seen to be valid for $|r - 1| < 1/\sqrt{2}$. If we allow r to approach $1 -$, the desired conclusion is a consequence of Theorem 10.2, Chapter IV.

We may describe the physical effect of the operator $e^{r\sqrt{\mathcal{D}}}$. We saw in Chapter IV that the temperature $u(x, t)$ of a semi-infinite insulated bar (which is initially at zero temperature and whose end at $x = 0$ is held at temperature $\varphi(t)$ as the time changes) is given by

$$u(x, t) = \int_0^t h(x, t - y)\varphi(y) \, dy.$$

The above calculations show that

$$e^{r\sqrt{\mathcal{D}}} u(x, t) = \int_0^t h(x - r, t - y)\varphi(y) \, dy = u(x - r, t)$$

so that this differential time-variable operator produces a translation in the space variable. Compare with earlier operational results:

$$e^{rD}u(x, t) = u(x + r, t), \qquad D = \frac{\partial}{\partial x},$$

$$e^{rD^2}u(x, t) = u(x, t + r).$$

8 TIME-VARIABLE SINGULARITIES

We have seen in Theorem 4.2, Chapter IV, that a temperature function defined by a Poisson integral is analytic in the time variable at least in the strip of definition, $0 < t < c$, though for specific examples such as $k(x, t) * e^{x^2}$ the region of analyticity includes points for which $t = 0$. Here we introduce conditions under which there will surely be a singularity at *every* point of the x-axis. This result will enable us to give a precise formula for the optimal value b in Theorem 4. We prove first a preliminary result, the full force of which we do not need here, but which is of independent interest. In what follows $s = \sigma + i\tau$ is a complex variable, as usual.

Theorem 8.1

$$\text{1. } f(s) \in A, \quad |s| < \rho,$$

$$\Leftrightarrow \quad f(\sigma) = \int_{-\infty}^{\infty} k(y, \sigma)\varphi(y)\, dy, \quad |\sigma| < \rho, \tag{1}$$

where $\varphi(-y) = \varphi(y)$ and $\varphi(y) \in \{2, 1/(4\rho)\}$.

Here we have used the notation of Definition 7, Chapter X, to indicate that φ is entire of growth $\{2, 1/(4\rho)\}$. Set

$$f(s) = \sum_{n=0}^{\infty} a_n s^n, \quad \overline{\lim_{n \to \infty}} |a_n|^{1/n} \leq \frac{1}{\rho},$$

$$\varphi(y) = \sum_{n=0}^{\infty} b_{2n} y^{2n}, \quad b_{2n} = \frac{a_n n!}{(2n)!}. \tag{2}$$

Recall that

$$\int_{-\infty}^{\infty} k(y, \sigma) y^{2n}\, dy = v_{2n}(0, \sigma) = \frac{(2n)!\, \sigma^n}{n!}.$$

If b_n is defined to be 0 when n is odd, then

$$\overline{\lim_{n \to \infty}} \frac{2n}{e} |b_n|^{2/n} = \overline{\lim_{n \to \infty}} \frac{4n}{e} |b_{2n}|^{1/n}$$

$$= \overline{\lim_{n \to \infty}} \frac{4n}{e} |a_n|^{1/n} \left[\frac{n!}{(2n)!} \right]^{1/n}$$

$$= \overline{\lim_{n \to \infty}} |a_n|^{1/n} \leq \frac{1}{\rho}. \tag{3}$$

The last step was accomplished by Stirling's formula. By 7(2), Chapter X, this inequality concerning the b_n characterizes functions $\varphi(y)$ of class $\{2, 1/(4\rho)\}$. That is, $f(s) \in A$ for $|s| < \rho$ if and only if $\varphi(y) \in \{2, 1/(4\rho)\}$.

Now direct substitution of series (2) in integral (1) gives

$$\int_{-\infty}^{\infty} k(y, \sigma) \sum_{n=0}^{\infty} b_n y^{2n}\, dy = \sum_{n=0}^{\infty} a_n \sigma^n = f(\sigma).$$

This term-by-term integration is valid since by (3) we have

$$\int_{-\infty}^{\infty} k(y, \sigma) \sum_{n=0}^{\infty} |b_n| y^{2n}\, dy = \sum_{n=0}^{\infty} |a_n| \sigma^n < \infty, \quad 0 < \sigma < \rho.$$

8. TIME-VARIABLE SINGULARITIES

The conclusion of the theorem now follows immediately.

Theorem 8.2

1. $\alpha(y) \in \uparrow$ is not absolutely continuous,
2. $u(x, s) = \int_{-\infty}^{\infty} k(x - y, s)\, d\alpha(y), \quad 0 < \sigma < c,$ (4)
3. $-\infty < x_0 < \infty$

$\Rightarrow$
 A. $u(x_0, s) \in A, \quad \mathrm{Re}(1/s) > 1/c,$
 B. $\notin A \quad \text{at} \quad s = 0.$

By Theorem 4.2, Chapter IV, $u(x_0, s)$ is analytic in the disk $\mathrm{Re}(1/s) > 1/t_0$ for every positive $t_0 < c$, from which Conclusion A is evident. Under the present restrictions on $\alpha(y)$, we are to prove that $u(x, t)$ has a singularity at *every* point on the x-axis. Since

$$u(x - x_0, t) = \int_{-\infty}^{\infty} k(x - y, t)\, d\alpha(y - x_0),$$

it is only necessary to consider the case $x_0 = 0$,

$$u(0, s) = \int_0^{\infty} k(y, s)\, d[\alpha(y) - \alpha(-y)]. \tag{5}$$

To prove B, we proceed by contradiction and assume $u(0, s) \in A$ at $s = 0$. Then by Theorem 8.1,

$$u(0, \sigma) = \int_0^{\infty} k(y, \sigma)\varphi(y)\, dy, \tag{6}$$

where $\varphi(y)$ is entire. Both integrals (5) and (6) become Laplace integrals if y^2 is replaced by a new variable. Hence we may apply the uniqueness theorem for such integrals to obtain

$$\alpha(y) - \alpha(-y) = \int_0^y \varphi(z)\, dz.$$

That is, $\alpha(y) - \alpha(-y)$ is certainly absolutely continuous. Since $\alpha(y)$ and $-\alpha(-y)$ are both nondecreasing, we can conclude that $\alpha(y)$ is also absolutely continuous. This follows from the very definition. Given ϵ, we can find δ such that

$$\sum_{k=1}^{n} [\alpha(y_k + h_k) - \alpha(y_k)] + \sum_{k=1}^{n} [\alpha(-y_k) - \alpha(-y_k - h_k)] \leq \epsilon \tag{7}$$

for every set of nonoverlapping intervals $(y_k, y_k + h_k)$ such that $\Sigma_1^n h_k \leq \delta$. Since each of the two sums (7) is ≥ 0, then each must be $\leq \epsilon$ separately so that $\alpha(y)$ is absolutely continuous. But this contradicts hypothesis 1, so that $u(0, s) \not\in A$ at $s = 0$.

Theorem 8.3

1. $\alpha(y) \in \uparrow \cdot B$ is not absolutely continuous,
2. $u(x, t) = \int_{-\infty}^{\infty} k(x - y, t) \, d\alpha(y), \qquad 0 < t < \infty,$
3. $0 < c$

$\Rightarrow$

A. $u(x, c - t) = \sum_{n=0}^{\infty} \frac{u^{(2n)}(x, c)(-t)^n}{n!},$

$$-\infty < x < \infty, \quad |t| < c, \qquad (8)$$

B. $\varlimsup_{n \to \infty} \frac{|u^{(2n)}(x, c)|^{1/n}}{n} = \frac{1}{ce}. \qquad (9)$

To the hypotheses of the previous theorem, we have added the boundedness of $\alpha(y)$, so that integral (4) converges for $0 < \sigma < \infty$. Hence for any positive c, $u(x, c - s)$ is an analytic function of s in the disk $|s| < c$, so that conclusion A is a consequence of Taylor's theorem. By Theorem 8.2, there is a singularity at $s = c$, so that the radius of convergence of the power series (8) is c. By Hadamard's formula

$$\varlimsup_{n \to \infty} \left| \frac{u^{(2n)}(x, c)}{n!} \right|^{1/n} = \frac{1}{c}.$$

By Stirling's formula this is equivalent to B.

Corollary 8.3

$u(x, c)$ is entire, order 2, type $\frac{1}{4c}$, $0 < c < \infty$.

By the known formula for the type of an entire function of order 2 [Boas 1954; 11] we must show that

$$\varlimsup_{n \to \infty} \frac{|u^{(n)}(x, c)|^{2/n}}{n} = \frac{1}{2ce}. \qquad (10)$$

This follows from (9) in so far as even values of n are concerned. We know that the Poisson integral may be differentiated under the integral sign, so that from (4)

$$u'(x, s) = \int_{-\infty}^{\infty} k'(x - y, s) \, d\alpha(y), \qquad 0 < \sigma < \infty.$$

Hence for fixed x, $u'(x, c - s) \in A$ at least for $|s| < c$. Hence,

$$u'(x, c - t) = \sum_{n=0}^{\infty} \frac{u^{(2n+1)}(x, c)(-t)^n}{n!},$$

and by Hadamard's formula

$$\overline{\lim_{n \to \infty}} \frac{|u^{(2n+1)}(x, c)|^{1/n}}{n} \leq \frac{1}{ce}. \tag{11}$$

This *inequality* combined with *equality* (9) yields equality (10), as desired. Unlike $u(x, s)$ defined by (4), the derivative (10) need not be singular at $t = 0$. For example, if $\alpha(y)$ is odd, then $u'(0, s) \equiv 0$, and the left-hand side of (12) is 0 for $x = 0$.

We show by examples that for the truth of Theorem 8.2 the hypotheses cannot be relaxed essentially. For example, if the condition $\alpha(y) \in \uparrow$ is dropped, there need be no singularity at $s = 0$. If $\alpha(y)$ is the step-function which is unity on $(-1, 1)$ and is zero elsewhere, $\alpha(y)$ is not monotonic. We have

$$u(x, t) = k(x + 1, t) - k(x - 1, t), \qquad u(0, t) \equiv 0.$$

The theorem fails for the single value $x = 0$.

For the validity of Theorem 8.2 it is also essential that $\alpha(y)$ should not be absolutely continuous for, if $\alpha(y) = y$, $u(x, t) \equiv 1$, and there are no singularities at all. Or, if $\alpha(y) = y$ for $y > 0$, $\alpha(y) = 0$ for $y \leq 0$, and $\alpha(y) \in \uparrow$ and is absolutely continuous. But

$$u(x, t) = \int_0^{\infty} k(x - y, t) \, dy = \frac{1}{2} \operatorname{erfc}\left(-\frac{x}{\sqrt{4t}}\right).$$

For every x except $x = 0$ this function has a singularity at $t = 0$. At $x = 0$ it is constant! Finally, consider the familiar integral

$$u(x, t) = \int_{-\infty}^{\infty} k(x - y, t) e^{-y^2} \, dy = \frac{e^{-x^2/(1+4t)}}{\sqrt{1 + 4t}}.$$

Again $\alpha(y) \in \uparrow$ and is absolutely continuous, but $u(x, t)$ is analytic at $t = 0$ for *every* fixed x.

As a consequence of Theorem 8.3, we may obtain a precise formula for determining the optimal value b of Theorem 4, if it is known that the optimal function $\alpha_b(y)$ is not absolutely continuous.

Theorem 8.4

1. $\alpha(y) \in \uparrow \cdot B\quad$ is not absolutely continuous,
2. $f(x) = \int_{-\infty}^{\infty} k(x - y, b)\, d\alpha(y), \quad b > 0,$ (12)

$\Rightarrow$
$$\frac{1}{b} = \overline{\lim_{n \to \infty}} \frac{e}{n} |f^{(2n)}(0)|^{1/n}.$$ (13)

This is because $f(x)$, as defined by (12), is $u(x, b)$ of Theorem 8.3. Conclusion B with $x = 0$ is (13).

As an illustration, consider the first example of §4,

$$f(x) = k(x, \tfrac{1}{4}) = \int_{-\infty}^{\infty} k(x - y, \tfrac{1}{4})\, dU(y)$$

$$= \frac{1}{2\pi} \int_{-\infty}^{\infty} e^{ixy - (y^2/4)}\, dy.$$

Here we have used Theorem 5, Chapter III. Then

$$f^{(2n)}(0) = \frac{(-1)^n}{2\pi} \int_0^{\infty} y^{2n} e^{-y^2/4}\, dy$$

$$= \frac{(-1)^n}{2\pi} \int_0^{\infty} z^{(2n-1)/2} e^{-z/2}\, dz$$

$$= (-1)^n \Gamma(n + \tfrac{1}{2}) 4^{(2n+1)/2}$$

$$= \frac{(-1)^n}{\sqrt{\pi}} \frac{(2n)!}{n!}$$

$$4 = \lim_{n \to \infty} \left(\frac{e}{n}\right) \left[\frac{(2n)!}{\sqrt{\pi}\, n!}\right]^{1/n}.$$

Hence $b = \tfrac{1}{4}$, as predicted by the theorem.

BIBLIOGRAPHY

Appell, P.
1892 Sur l'équation $(\partial^2 z/\partial x^2) - (\partial z/\partial y) = 0$ et la théorie de la chaleur. *J. Math. Pures Appl.* **8**, 187–216.

Banach, S.
1932 "Theorie des Opérations Lineaires." Garasinski, Warsaw.

Bellman, R.
1961 "A Brief Introduction to Theta Functions." Holt, New York.

Bilodeau, G. G.
1974 On generalized heat polynomials. *SIAM J. Math. Anal.* **5**, 43–50.

Birkhoff, G., and Kotik, J.
1954 A note on the heat equation. *Proc. Amer. Math. Soc.* **50**, 162–167.

Blackman, J.
1952 The inversion of solutions of the heat equation for the infinite rod. *Duke Math. J.* **19**, 671–682.

Blackman, J., and Pollard, H.
1959 The finite convolution transform. *Trans. Amer. Math. Soc.* **75**, 399–409.

Boas, R. P.
1954 "Entire Functions." Academic Press, New York.

Bochner, S.
1959 "Lectures on Fourier Integrals" (Ann. Math. Stud., No. 42). Princeton Univ. Press, Princeton, New Jersey.

Bragg, L. R.
1965 The radial heat polynomials and related functions. *Trans. Amer. Math. Soc.* **119**, 270–290.

Carslaw, H. S., and Jaeger, J. C.
1948 "Conduction of Heat in Solids." Oxford Univ. Press (Clarendon), London and New York.

Cooper, J. L. B.
1954 The uniqueness of the solution of the equation of heat conduction. *J. London Math. Soc.* **25**, 173–180.

Diaz, J. B., and Curtis, S. M.
1973 An initial value problem for a class of higher order partial differential Equations related to the heat equation. *Ann. Mat. Pura App.* **97**, 115–187.

Doetsch, G.
1936 Les équations aux dérivées partielles du type parabolique. *Enseignement Math.* **35**, 43–87.

Ehrenpreis, L.
1961 Analytically uniform spaces and some applications. *Trans. Amer. Math. Soc.* **101**, 47–52.
1970 "Fourier Analysis in Several Complex Variables." Wiley, New York.

Erdélyi, A.
1953 "Higher Transcendental Functions," Vol. 2 (Bateman Manuscript Project). McGraw-Hill, New York.
1954 "Tables of Integral Transforms," Vol. 1 (Bateman Manuscript Project). McGraw-Hill, New York.

Fourier, J.
1878 "Analytic Theory of Heat." Cambridge University Press, Cambridge.

Gehring, F. W.
1960 The boundary behavior and uniqueness of solutions of the heat equation. *Trans. Amer. Math. Soc.* **94**, 337–364.

Gevrey, M.
1913 Sur les équations aux dérivées parielles du type parabolique. *J. Math. Pures Appl.* **9**, 305–471.

Goursat, E.
1923 "Cours d'Analyse Mathématique," Vol. 3, Chapter 29. Gauthier-Villars, Paris.

Haimo, D. T.
1965 Functions with the Huygens property. *Bull. Amer. Math. Soc.* **71**, 528–532.
1966a Expansions in terms of generalized heat polynomials and of their Appell transforms. *J. Math. Mech.* **15**, 735–758.
1966b Generalized temperature functions. *Duke Math. J.* **33**, 305–322.
1967 Series representations of generalized temperature functions. *SIAM J. Appl. Math.* **15**, 359–367.
1970 Series expansions and integral representations of generalized temperatures. *Illinois J. Math.* **14**, 621–629.
1973 Widder temperature representations. *J. Math. Anal. Appl.* **41**, 170–178.

Haimo, D. T., and Cholewinski, F. M.
1966 Integral representations of solutions of the generalized heat equation. *Illinois J. Math.* **10**, 623–638.
1967 Laguerre temperatures. Proc. Conf. *Southern Illinois Univ., Edwardsville, April 27–29, 1967* pp. 197–226.

Hartman, P., and Wintner A.
1950 On the solutions of the equation of heat conduction. *Amer. J. Math.* **72**, 367–395.

Hellwig, G.
1964 "Partial Differential Equations" Part 1, §4, and Part 3, §2. Ginn (Blaisdell), Boston, Massachusetts.

Hirschman, I. I.
1952 A note on the heat equation. *Duke Math. J.* **19**, 487–492.

Hirschman, I. I., and Widder, D. V.
1955 "The Convolution Transform" Princeton Univ. Press, Princeton, New Jersey.

Holmgren, E.
1924 Sur les solutions quasi-analytiques de l'équation de la chaleur. *Ark. Mat.* **18** (9), 1–9.
Kampé de Fériet, J.
1959 Heat equation and Hermite polynomials. *The Golden Jubilee Commemoration Volume (1958–1959)*, Calcutta Math. Soc.
Lévy, P.
1932 Sur un problème de calcul des probabilités lié a celui du refroidissement d'une barre homogène. *Ann. Scuola Norm. Sup. Pisa* **1**, 283–296.
Pollard, H.
1944 One sided boundedness as a condition for the unique solution of certain heat equations. *Duke Math. J.* **11**, 651–653.
1953 Distribution functions containing a Gaussian factor. *Proc. Amer. Math. Soc.* **4**, 578–583.
Pollard, H., and Widder, D. V.
1969 Gaussian representations related to heat conduction. *Arch. Rational Mech. Anal.* **35**, 253–258.
1970 Inversion of a convolution transform related to heat conduction. *SIAM J. Math. Anal.* **1**, 527–532.
1960 Certain solutions of the heat conduction equation. *Quart. Appl. Math.* **18**, 97–106.
Rosenbloom, P. C., and Widder, D. V.
1958 A temperature function which vanishes initially. *Amer. Math. Monthly* **65**, 607–609.
1959 Expansions in terms of heat polynomials and associated functions. *Trans. Amer. Math. Soc.* **92**, 220–226.
Täcklind, S.
1936 Sur les classes quasianalytiques des solutions des équations aux dérivées partielles du type parabolique. *Nova Acta Regiae Soc. Sci. Upsal.* **10**, 1–57.
Titchmarsh, E. C.
1939 "The Theory of Functions." Oxford Univ. Press (Clarendon), London and New York.
Tychonov, A. N.
1935 Théorèmes d'unicité pour l'equation de la chaleur. *Mate. Sb.* **42**, 199–216.
Tychonov, A. N., and Samarski, A. A.
1964 "Partial Differential Equations of Mathematical Physics," Vol. 1, Chapter 3. Holden-Day, San Francisco, California.
Whittaker, E. T., and Watson, G. N.
1943 "Modern Analysis." Macmillan, New York.
Widder, D. V.
1930 Singular points of functions which satisfy the partial differential equation of the flow of heat. *Bull. Amer. Math. Soc.* **36**, 687–694.
1944 Positive temperatures on an infinite rod. *Trans. Amer. Math. Soc.* **55**, 85–95.
1946 "The Laplace Transform." Princeton Univ. Press, Princeton, New Jersey.
1951a Necessary and sufficient conditions for the representation of a function by a Weierstrass transform. *Trans. Amer. Math. Soc.* **71**, 430–439.
1951b Weierstrass transforms of positive functions. *Proc. Nat. Acad. Sci.* **37**, 315–317.
1953 Positive temperatures on a semi-infinite rod. *Trans. Amer. Math. Soc.* **75**, 510–525.
1955 The heat equation and the Weierstrass transform. *Proc. Conf. Differential Equations Univ. Maryland, March 17–19, 1955* pp. 227–234.
1956 Integral transforms related to heat conduction. *Ann. Mat. Pura Appl.* **42**, 279–305.
1961a Transformations associées à l'équation de la chaleur. *C. R. Acad. Sci. Paris* **253**, 915–917.
1961b Series expansions of solutions of the heat equation in n dimensions. *Ann. Mat. Pura Appl.* **55**, 389–410.

1961c "Advanced Calculus." Prentice-Hall, Englewood Cliffs, New Jersey.
1962a Series expansions in terms of the temperature functions of Poritsky and Powell. *Quart. Appl. Math.* **20**, 41–47.
1962b Analytic solutions of the heat equation. *Duke Math. J.* **29**, 497–504.
1963a Sur une classe de fonctions $u(x, y, t)$ harmoniques en (x, y) et satisfaisant l'équation de la chaleur en (x, t). *C. R. Acad. Sci. Paris* **256**, 2751–2753.
1963b Positive solutions of the heat equation. *Bull. Amer. Math. Soc.* **69**, 111–112.
1963c The role of the Appell transformation in the theory of heat conduction. *Trans. Amer. Math. Soc.* **109**, 121–134.
1963d Functions of three variables which satisfy both the heat equation and Laplace's equation in two variables. *J. Austral. Math. Soc.* **3**, 396–407.
1964a The inversion of a transform related to the Laplace transform and to heat conduction. *J. Austral. Math. Soc.* **4**, 1–14.
1964b A problem of Kampé de Fériet. *J. Math. Anal. Appl.* **9**, 458–467.
1966 Some analogies from classical analysis in the theory of heat conduction. *Arch. Rational Mech. Anal.* **21**, 108–119.
1967a Inversion of a heat transform by use of series. *J. Analyse Math.* **18**, 389–413.
1967b Expansions in terms of the homogeneous solutions of the heat equation. *Proc. Southern Illinois Univ., Edwardsville, April 27–29, 1967* pp. 171–196.
1968 Homogeneous solutions of the heat equation. *Proc. Bloomington Conference, 1968* pp. 379–398.
1969 Expansions in series of homogeneous temperature functions of the first and second kinds. *Duke Math. J.* **36**, 495–510.
1971 "An Introduction to Transform Theory." Academic Press, New York.
1972 Time-variable singularities for solutions of the heat equation. *Proc. Amer. Math. Soc.* **32**, 209–214.

INDEX

A

Abelian theorem, 64
Absolutely continuous, 255
Absorption postulate, 2
Addition formula for $k(x, t)$, 32, 33, 37
Adjoint, 7
Adjoint heat equation, 15
Affine transformation, 12
Analogies, 195
Appell, P., 14, 229
Appell's problem, 229
Appell transformation, 13, 53, 182, 197, 216, 230, 235, 240
  in higher dimensions, 209
Askey, R. A., 94
Associated function of heat polynomial, 165–168
  asymptotic estimates for, 172
  bound on, 173
  criteria for expansion in, 201–203
  in higher dimensions, 209
  region of convergence of series of, 182
  series of, 200, 201, 203
Asymptotic estimates
  for associated functions, 172
  for heat polynomials, 171, 174
  for Hermite polynomials, 174

B

Banach, S., 122
Bilaterial convolution, 32
Biorthogonal relation, 170, 199
Blackman, J., 159
Blackman's example, 159
Boas, R. P., 243, 256
Bochner, S., 237
Boundary-value problem, 17
Bounded variation, 60
BTU, 1, 2

C

Calorie, 2
Carslaw, H. S., 204
Cauchy criterion, 63
Cauchy integral formula, 46, 77, 172
Cauchy problem, 6, 7, 18, 27, 32, 43, 50, 52, 53, 57, 60, 156
Characteristic curve, 6, 7
Classes, relation among, 49

C

Class A, 45
Class I, 44
  series expansion in, 50
Class II, 44
  series expansion in, 46, 52
Class III, 45
  series expansion in, 53
Compactness, weak, 122, 132
Complementary error function, $l(x, t)$, 10
Conduction postulate, 2
Conditions
  Dirichlet, 90, 100
  Jordan, 100
Continued fraction for homogeneous function, 223–225
Convergence of Poisson transform, 62
Convolution, 13
Criterion for temperature function, 23

D

Dedekind cut, 248
Density, 4
Derived source solution, $h(x, t)$, 10, 39
Derived theta function, $\varphi(x, t)$, 86
Diffusivity, 5
Dirichlet conditions, 90, 100
Dirichlet series, 103

E

Elliptic theta functions, $\varphi(x, t)$, $\theta(x, t)$, 10, 86
Entire function
  growth of, 185
  type of, 243, 256
Erdélyi, A., 57, 174
Euler equation, 222
Examples, 191–194, 214, 215, 232–234, 244–246
Exponential solutions, 8

F

Finite rod
  bounded temperature on, 130
  positive temperature on, 148–151
Flux, 18
Fourier coefficients, 25, 91
Fourier, J., 100
Fourier ring, 100
Fourier series, 25, 90, 96, 100, 104
Fourier transform, 247
Frequency function, 241
Fubini theorem, 157, 219, 239, 241

G

Gaussian representation, 241, 242
Gehring, F. W., 156
Generating function, 175, 200
  for associated function, 169
  for heat polynomial, 169
  in higher dimensions, 209, 210
Green's function
  property of, 112
  definition of, 107
  existence of, 108
  Problem I by, 110
  for rectangle, 107, 113
  representation by, 109
Green's theorem, 7, 8, 20, 83, 117, 206
Growth of entire function, 185

H

$H^\Delta \neq H^o$, 163
$H^o \neq H^\Delta$, 163
Hadamard formula, 180
Haimo, D. T., 194
Hartman, P., 151
Heat equation, 3, 207
Heat polynomials, 8, 197
  associated function of, 165–168, 198
  bound on, 173
  series of, 177
Heine–Borel theorem, 22, 45
Helly, E., 132
Hermite differential equation, 34
Hermite polynomial, 57, 170, 174
Higher dimensions, 204–215
  Appell transform in, 209
  associated function in, 209
  generating function in, 209, 210
  heat polynomial in, 208
  Huygens property in, 208
  series expansion in, 210–213
  source solution in, 208
Hirschman, I. I., 237
h-Lebesgue transform
  inversion of, 78
Homogeneity of $k(x, t)$, 34

INDEX 265

Homogeneous temperature function, 34, 216
   continuted fraction for, 223–225
   decomposition of, 226
   degree of, 216
   examples of, 220
   of first kind, 217, 230
   of second kind, 217, 231, 232
   recurrence relations for, 222
   table of, 227
   totality of, 218, 220
   totality of, in H^o, 221
h-Stieltjes transform, 70
   in Class II, 76
   inversion of, 79
h-transform, 70
   alternate inversion of, 249–253
   analytic, 75
   in H, 74
Huygens, C., 156
Huygens property, 155, 156, 183, 198
   in higher dimensions, 208
$h(x, t)$, derived source solution, 10
   Laplace transform of, 39

I

Infinite rod
   bounded temperature on, 122–126
   positive temperature on, 132
   Stieltjes integral representation on, 136
   uniqueness on, 133
Integral representation
   basic, 82
   of $k(x, t)$, 35
Integral transform, 60
Isolated singularities of temperature function, 116

J

Jacobi, C. G. J., 86, 100
Jager, J. C., 204
Jordan conditions, 100

K

k-transform
   in H, 80
   in I, 81
   in II, 81
   inversion of, 80

$k(x, t)$, source solution, 10, 30, 31
   addition formula for, 32, 33, 37
   bounds on, 87
   characterization of, 35
   homogeneity of, 34
   integral representation of, 35
   Laplace transform of, 38

L

Laplace asymptotic method, 66
Laplace transform, 63–65, 75, 160, 218, 236, 237
   of $h(x, t)$, 39
   of $k(x, t)$, 38
   of $l(x, t)$, 40
   of $\varphi(x, t)$, 151
   uniqueness theorem for, 255
Laurent series
   criteria for expansion in, 202
Lebesgue integrable, 60
Lebesgue limit theorem, 61, 72, 75
L'Hospital rule, 177
Liouville theorem, 236
Lord Kelvin, see Thomson, W.
$l(x, t)$, complementary error function, 10, 40, 219

M

Maclaurin theorem, 45, 54, 188, 213
Maximum principle, 20, 21
Maxwell, J. C., 5
Methods for generating solutions, 10

N

Newton law of radiation, 18
Null solutions, 27, 156, 221

O

Operational calculus, 41

P

φ-transform
   positiveness of kernel for, 92
$\varphi(x, t)$, derived theta function, 10, 86
   alternate expansion of, 90
   analyticity of, 88
   periodicity of, 88

Physical model, 1
Poisson–Lebesgue transform, 60
  inversion of, 65
Poisson–Stieltjes transform, 60
  inversion of, 68
Poisson transform, 32, 56, 60, 156, 229, 253
  θ-transform as, 94
  analytic, 64
  of bounded function, 122
  conditionally convergent, 161
  convergence of, 62
  in H, 64
  representing positive temperature function, 132, 136
Pollard, H., 242
Polynomial series, 200
  criteria for expansion in, 201
  radius of convergence of, 180
  region of convergence of, 177
  strip of convergence of, 180
Positive definite function, 237
Positive temperature function, 132, 136, 140–147, 239
Postulate A, 2
  alternate form of, 205
  for solids, 206
Postulate B, 2
  alternate form of, 205
  for solids, 206
Prime, 87
Principle of reflection, 26, 115, 116
Problem I, 18, 96
  by Green's function, 110
  modified, 19
  solution of, 24, 101, 102
  unique solution of, 104
  uniqueness for finite rod, 22

R

Radius of convergence of polynomial series, 180
Reflection principle, 29, 115 (Schwarz), 116
Region of convergence
  of associated series, 182
  of polynomial series, 180
Riemann–Lebesgue theorem, 180
Rosenbloom, P. C., 27

S

Schwarz reflection principle, 115
Selection principle, 122, 132
Semi-infinite rod
  bounded temperature on, 124–129
  positive temperature on, 140–147
  uniqueness for, 29, 116, 137–140
Series expansions
  in Class I, 50
  in Class II, 52
  in Class III, 53
  in higher dimensions, 210–213
  of temperature functions, 169
Series of heat polynomials, 117, 227
  criteria for expansion in, 228
Series of temperature functions, 114
Singularities
  conditions for time-variable, 255
  of temperature function, 116
  time-variable, 253–258
Sink, 112
Solution of Problem I, 24
Source, 112
  strength of, 31
Source solution, $k(x, t)$, 10, 30
  addition formula for, 32, 33, 37
  characterization of, 35
  derivatives unbounded, 120
  derived, 10, 39
  in higher dimensions, 208
  integral representation of, 35
  Laplace transform of, 38, 40
  properties of, 31
Specific heat, 2
Statistical problem, 241–249, 258
Sterling formula, 174, 178
Stieltjes integral, 142, 147
Stieltjes integral representation
  on finite rod, 148
  on infinite rod, 136
  on semi-infinite rod, 143
Strip of convergence
  of polynomial series, 180
Superscript, 87

T

Taylor's formula, 49

Temperature function, 15
  analytic character of, 84
  bounded, 120, 122
  bounded below, 120
  bounded on finite rod, 130
  bounded on infinite rod, 123
  bounded on semi-infinite rod, 124–129
  further classes of 153
  homogeneous, 216–234
  isolated singularities of, 116
  not entire, 58
  positive, 132, 136, 235–241
  series expansion of, 114, 169, 183, 186, 189
  in solids, 207
Thermal conductivity, 2
Thermometric conductivity, 5
θ-transform
  in H, 94, 97
  inversion of, 94, 99
  as Poisson transform, 94
  positiveness of kernel for, 92
$\theta(x, t)$, elliptic theta function, 10, 86, 100
  alternate expansion of, 90
  in H, 89
  periodicity of, 88

Thomson, W., (Lord Kelvin), 5
Time-variable singularities, 253–258
  conditions for, 255
Titchmarsh, E. C., 53, 55, 61, 68, 72, 96, 152
Translation, 12

U

Unilateral convolution, 32
Uniqueness, 18
  for finite rod, 22
  for infinite rod, 26, 28, 133
  for semi-infinite rod, 29, 116, 137–140
Unit of heat, 2
Unit of temperature, 2

V

Venn diagram, 50

W

Weak compactness, 122, 132
Weierstrass M-test, 51, 54
Weierstrass theorem, 26, 89
Wintner, A., 151